T0179612

Mathematical Modelling of
System Resilience

RIVER PUBLISHERS SERIES IN MATHEMATICAL AND ENGINEERING SCIENCES

Series Editors

TADASHI DOHI
Hiroshima University
Japan

ALIAKBAR MONTAZER HAGHIGHI
Prairie View Texas A&M University
USA

MANGEY RAM
Graphic Era University
India

Indexing: All books published in this series are submitted to the Web of Science Book Citation Index (BkCI), to SCOPUS, to CrossRef and to Google Scholar for evaluation and indexing.

Mathematics is the basis of all disciplines in science and engineering. Especially applied mathematics has become complementary to every branch of engineering sciences. The purpose of this book series is to present novel results in emerging research topics on engineering sciences, as well as to summarize existing research. It engrosses mathematicians, statisticians, scientists and engineers in a comprehensive range of research fields with different objectives and skills, such as differential equations, finite element method, algorithms, discrete mathematics, numerical simulation, machine leaning, probability and statistics, fuzzy theory, etc.

Books published in the series include professional research monographs, edited volumes, conference proceedings, handbooks and textbooks, which provide new insights for researchers, specialists in industry, and graduate students.

Topics covered in the series include, but are not limited to:

- Advanced mechatronics and robotics
- Artificial intelligence
- Automotive systems
- Discrete mathematics and computation
- Fault diagnosis and fault tolerance
- Finite element methods
- Fuzzy and possibility theory
- Industrial automation, process control and networked control systems
- Intelligent control systems
- Neural computing and machine learning
- Operations research and management science
- Optimization and algorithms
- Queueing systems
- Reliability, maintenance and safety for complex systems
- Resilience
- Stochastic modelling and statistical inference
- Supply chain management
- System engineering, control and monitoring
- Tele robotics, human computer interaction, human-robot interaction

For a list of other books in this series, visit www.riverpublishers.com

Mathematical Modelling of System Resilience

Kanchan Das

East Carolina University
USA

Mangey Ram

Graphic Era University, Dehradun
India

LONDON AND NEW YORK

Published 2019 by River Publishers
River Publishers
Alsbjergvej 10, 9260 Gistrup, Denmark
www.riverpublishers.com

Distributed exclusively by Routledge
4 Park Square, Milton Park, Abingdon, Oxon OX14 4RN
605 Third Avenue, New York, NY 10017, USA

Mathematical Modelling of System Resilience / by Kanchan Das, Mangey Ram.

Routledge is an imprint of the Taylor & Francis Group, an informa business

ISBN 978-87-7022-070-5 (print)

While every effort is made to provide dependable information, the publisher, authors, and editors cannot be held responsible for any errors or omissions.

Contents

6 Performance Measures of a Complex System with Possible Online Repair   **143**

Beena Nailwal, Bhagawati Prasad Joshi and Suraj Bhan Singh

7 Small Quadrotor Functioning under Rework Analysis   **171**

Nupur Goyal, Ayush Kumar Dua, Akshay Bhardwaj,
Akshay Kumar and Mangey Ram

Preface

Resilience concepts are often complex yet the importance of system resilience for the persistence of the performance of a system is accepted by every organization including business, engineering systems, academic institution and government organizations. The National Academy of Science(NRC) [1] defined resilience as the ability to plan and prepare for, absorb, recover from and adapt to adverse events. Since each organization or system is exposed to some external or internal risks or disruptions, resilience development is given importance by every organization and systems. For understanding and creating system resilience mathematical modeling-based approaches are the most common and established ones. This book Mathematical Modelling of System Resilience will include various modelling techniques and approaches for understanding and creating resilience to overcome/face/manage risks and disruptions in supply chain process; in engineering and technical systems, and communities. The book will benefit engineers and other practitioners who are involved in systems resilience development for their own systems by providing some innovative alternative ways to choose from or consult for the approaches to resilience. It may help them make their system more resilient to risks and disruptions from external and internal sources in a cost-effective way.

Chapter 1 includes a classic background of Supply Chain Risks and their relationship and transformation to vulnerability and disruptions supported by extensive empirical research from the literature. The author elegantly and nicely established the requirements and importance for supply chain resilience to motivate academics and supply chain managers by linking consequences of supply chain risks, vulnerability and the disruptions through the empirical research-based inferences. The chapter also addressed various concepts of resilience including resilience in the perspectives of supply chain disruptions with severe consequences. It considers all these with the objective of creating importance for supply continuity. Based on examples of various disruptions and definition for resilience the author of the Chapter 1 argued "while resilience is needed in these situations, recovery and return to stability may be very challenging amidst complete devastation. Supply continuity planning and resilience will help to address this type of risk but will also address a much wider range of risks as well". The chapter proposed a schematic model by including a basic supply chain process; internal and external integration; and layers of resilience factors including flexibility, agility, adaptability, visibility, and Strategic Sourcing issues.

Chapter 2 presents the mathematical modeling-based systems resilience approach for designing a resilient consumer product supply chain. The design framework plans resilience creation for containing failures and disruption by examining and understanding types, frequencies and intensities of risks and disruptions the system has been facing. The research also formulated resilience performance metrics such that supply chain management may plan risk and disruption containment measures to achieve their expected level of resilience. It also includes a numerical example to guide interested readers to plan and develop resilience for their own unique systems.

Chapter 3 presents an overview of resilience including its emergence of the resilience concepts. It presents different approaches to resilience including engineering, sociological and ecological. The chapter also presented the mathematical aspects of resilience. In the mathematical model-based approach of resilience, the author defined model for the dynamic system, stable system and changes in a system that happens naturally and changes from external disturbances starting from a stage due to some change reasons. The chapter also presents a mathematical model for ecological resilience. Based on Holling and Crawford's seminal research of 1996 [2] the chapter defined resilience in the perspectives of Ecology as the magnitude of disturbance that can be absorbed before the system changes its structure. The chapter also considered and discussed the four important aspects of ecological resilience. It defined resilience based on the four ecological aspects as the capacity of a system to absorb disturbance and reorganize while undergoing a change so as to still retain essentially the same function, structure, identity, and feedbacks. In this definition latitude, resistance, precariousness and panarchy are also introduced. Latitude of resilience refers to the maximal intensity of disruption the system can take without losing its ability to recover. This strongly reminds of the concept of ecological resilience, where the magnitude of disturbance before changing the structure of the system was considered.

Chapter 4 presented a quantitative evaluation methodology for the resiliency of electric traction drive as a typical case for similar systems. Such traction drive includes multilevel electric energy inverter and multiphase electric motor. The chapter suggested considering such electric traction drive as a case of propulsion system with select few states and degraded values of performance. The authors of the chapter also suggested using a degree of resiliency criteria in estimating resiliency of such complex system. For safety critical system the chapter recommended that the vehicle traction electric drive should have the possibility to operate in every degraded state after occurrences of several failures of electric traction drive components. As such for practical realization of such requirements, all units and subsystem of traction drive should be resilient. The chapter defined the degree of resiliency (DOR) model as the capacity of safety critical system to support the required level of performance in the case of a failure occurring in its units. The DOR model is the ratio of performance at the nominal and reduced level, multiplied by the relevant time intervals. The chapter provided example trends for various DOR applicable to the multilevel inverter. The authors also studied and presented graphically DOR values for a number of critical failures. In addition, the chapter included multi-state system reliability Markov models including transition probability-based analysis and examples of resilient traction drive for safety critical failures analysis in the case of multi-phase motors.

Chapter 5 presents a method for assessing climate related hydrological risks based on satellite observations and ground measurement data. It includes a mathematical model for expansion of dangerous process for analyzing data and defining its spectral indicators. Although the model and overall system approach are developed for a specific region, the approach may be applied to any region.

Chapter 6 presents the performance measure of a complex system with the possibility of on-line repair for the persistence of operation, which is always the target for creating resilience. The chapter evaluated performance by using the Supplementary variable technique, Laplace transformation and copula methodology. The chapter considered reliability model through a complex system in which two subsystems A and B are arranged in series. Subsystem A is circular consecutive 2-out of-3: F system and Subsystem B is circular consecutive 1-out of-3: P and 2-out of-3: F system. The chapter formulated mathematical model to represent the system and solved the model taking Laplace's transformation and presented a numerical example. The chapter presented the analysis and findings by various graphs.

Chapter 7 presents a stochastic programming model for the measures of the reliability of a vertical take-off and Landing (VTOL) prototype system. The chapter considers several economic characteristics of the reliability of a repairable VTOL system that are important for its performance. The authors of the chapter formulated a Markov chain-based model for the system and evaluated various reliability measures including system reliability and availability by using supplementary variable technique, Laplace and inverse Laplace transformation. The chapter solved a numerical example and presented their availability and reliability function in graphs. The authors also conducted sensitivity analysis and presented reliability sensitivity with time. Such model and analysis presented in the research will facilitate users to make the similar system (like helicopter drive system) robust and resilient.

[1] NRC (National Research Council) (2012). Disaster resilience: a national imperative. The National Academies Press, Washington.
[2] Holling, C. S., and Crawford, S. (1996). Engineering resilience versus ecological resilience. In *Engineering within ecological constraints,* 32.

Acknowledgments

The editors acknowledge River Publishers for this opportunity and professional support. Also, we would like to thank all the chapter authors and reviewers for their availability for this work.

List of Contributors

Akshay Bhardwaj, *Department of Computer Science and Engineering, Graphic Era University, Dehradun 248002, India*

Akshay Kumar, *Tula's Institute, The Engineering and Management College, Uttarakhand, India; E-mail: akshaykr1001@gmail.com*

Ayush Kumar Dua, *Department of Computer Science and Engineering, Graphic Era University, Dehradun 248002, India;*
E-mail: duaayush5@gmail.com

Beena Nailwal, *G. B. Pant University of Agriculture and Technology, Pantnagar; E-mail: bn4jan@gmail.com*

Benni Thiebes, *1. German Committee for Disaster Reduction DKKV, Bonn 53113, Germany*
2. COMTESSA, Universität der Bundeswehr München, Neubiberg 85577, Germany; E-mail: benni.thiebes@dkkv.org

Bhagawati Prasad Joshi, *Seemant Institute of Technology, Pithoragarh, India; E-mail: bpjoshi.13march@gmail.com*

Hans-Georg Herzog, *Institute of Energy Conversion Technology, Technical University of Munich, Munich, Germany; E-mail: hg.herzog@tum.de*

Igor Artemenko, *Scientific Centre for Aerospace Research of the Earth, National Academy of Sciences of Ukraine;*
E-mail: igor.artemenko@casre.kiev.ua

Igor Bolvashenkov, *Institute of Energy Conversion Technology, Technical University of Munich, Munich, Germany; E-mail: igor.bolvashenkov@tum.de*

Ilia Frenkel, *Center for Reliability and Risk Management, SCE – Shamoon College of Engineering, Beer-Sheva, Israel; E-mail: iliaf@frenkel-online.com*

Ivan Kopachevsky, *Scientific Centre for Aerospace Research of the Earth, National Academy of Sciences of Ukraine; E-mail: ivankm@ukr.net*

Jaqueline Hemmers, *1. German Committee for Disaster Reduction DKKV, Bonn 53113, Germany*
2. COMTESSA, Universität der Bundeswehr München, Neubiberg 85577, Germany; E-mail: jaqueline.hemmers@dkkv.org

Kanchan Das, *College of Engineering and Technology, East Carolina University, NC 27858, USA; E-mail: DASK@ecu.edu*

Mangey Ram, *Department of Mathematics, Graphic Era University, Dehradun, India; E-mail: mangeyram@gmail.com*

Markus Gerschberger, *LOGISTIKUM, University of Applied Sciences Upper Austria, Steyr 4400, Austria;*
E-mail: Markus.Gerschberger@fh-steyr.at

Matthias Winter, *LOGISTIKUM, University of Applied Sciences Upper Austria, Steyr 4400, Austria; E-mail: Matthias.Winter@fh-steyr.at*

Maxim Yuschenko, *Scientific Centre for Aerospace Research of the Earth, National Academy of Sciences of Ukraine;*
E-mail: max.v.yuschenko@gmail.com

Nupur Goyal, *Department of Mathematics, Garg P. G. College, Laksar, Uttarakhand, India; E-mail: nupurgoyalgeu@gmail.com*

Richard Monroe, *Longwood University, Farmville, VA 23909, USA, E-mail: ricklager85@yahoo.com*

Stefan Pickl, *1. German Committee for Disaster Reduction DKKV, Bonn 53113, Germany*
2. COMTESSA, Universität der Bundeswehr München, Neubiberg 85577, Germany; E-mail: stefan.pickl@unibw.de

Suraj Bhan Singh, *G. B. Pant University of Agriculture and Technology, Pantnagar; E-mail: drsurajbsingh@yahoo.com*

Yuriy V. Kostyuchenko, *Scientific Centre for Aerospace Research of the Earth, National Academy of Sciences of Ukraine;*
E-mail: yuriy.v.kostyuchenko@gmail.com

List of Figures

List of Tables

List of Abbreviations

CASRE	Scientific Center for Aerospace Research of the Earth of National Academy of Ukraine
CHB	multilevel cascaded H-bridge inverter
DOR	degree of resiliency
EOS	Earth Observation Satellite
ETM	Enhanced Thematic Mapper
EVI	Enhanced Vegetation Index
MF	motor failure
MODIS	The Moderate Resolution Imaging Spectroradiometer
MSS	Multi-Spectral Scanner
MSSR	MM multistate system reliability Markov model
NDVI	Normalized Difference Vegetation Index
NDWI	Normalized Difference Water Index
NOAA	The National Oceanic and Atmospheric Administration
PhF	phase failure
PRI	Photochemical Reflectance Index
PSI	Plant Stress Index
PSM	permanent magnet synchronous motor
SIPI	Structure Intensive Pigment Index
SRI	Spectral Reflectance Index
TM	Thematic Mapper
USAF	The United States Air Force
USGS	The United States Geological Survey
WMO	World Meteorological Organization

1

Developing Resilience in Supply Management

Richard Monroe

Longwood University, Farmville, VA 23909, USA
E-mail: ricklager85@yahoo.com

Supply chain risk management has developed as a research stream to address a wide range of vulnerabilities that companies face today. Risks must be addressed to ensure supply continuity and to maintain the operations that are needed to sustain company revenue and profitability. When risk events do occur, some measure of resilience is needed to mitigate the impact and to return to normal operating conditions quickly. The first aim of this chapter is to present a substantive sample of the supply chain risk management research to provide the context and the motivation for a systematic resilience program. A second aim is to present a multidimensional model that includes risk, integration and resilience. The proposed model can serve as a guide for organizations seeking to manage supply chain risks effectively by integrating supply chain processes and by developing resilience capabilities.

1.1 Introduction

Supply chain management depends on multidisciplinary coordination across all functions and that coordination must span across the internal/external boundary. The need for extensive coordination has led to a keen interest in supply chain integration among academics and practitioners. Over the last few decades, global supply chains have evolved and adapted to an array of factors which have added complexity and have interfered with integration efforts. Risk management, sustainability, and information security are among the major trends that have altered the supply chain landscape. Supply chain risk management (SCRM) has also garnered a lot of attention as a number of significant disruptions have impacted several industries. In the remainder of this section, some of the basic terminology will be presented prior to further analysis of supply chain risk.

As a foundation, it is important to understand the nature of supply chains and supply chain management. One description of a supply chain or supply network states that it is composed of "different entities that are connected by the physical flow of materials" (Craighead et al., 2007). While the material flow is the dominant flow when designing a supply chain, other flows are essential components of a supply chain as well. Information flow and financial flow are the other primary flows that make up the comprehensive set of supply chain flows. The following definition for "supply chain management" (SCM) offers a suitable description of the critical elements: "Supply chain management is the integration of key business processes from end user through original suppliers that provides products, services, and information that add value for customers and other stakeholders" (Lambert et al., 1998).

From these descriptions, one might surmise that significant risks may be associated with the material and information flows in a supply chain. Natural disasters and manmade disasters are two such examples of potential disruption risks. The financial flow is also subject to risk which may take many forms such as currency-exchange rate fluctuation, credit worthiness of supply chain partners, and other issues that may slow or interrupt payment transactions.

In the following section, some of the leading research about SCRM is discussed. Related topics are also introduced to develop the motivation for this chapter.

1.2 Supply Chain Risk Management

Common sources of supply chain risk addressed in previous research include: location, logistics, order processing, purchasing, quality, supply lead time, supply availability, and demand (Baramichai et al., 2009; Christopher and Lee, 2004; Larson and Gray, 2018). Research by Tang (2006) explicitly added uncertainty as a descriptor along with the following factors as sources of supply chain risk: uncertain demand; uncertain supply yields, uncertain lead times, uncertain supply capacity, uncertain supply cost, and uncertain price. Tang also presents a series of robust strategies to minimize the impact of these risks which will be discussed further in a later section (Tang, 2006).

Another example in this research stream suggests that risks take many forms including: financial, chaos, decision, and market risks (Christopher and Lee, 2004). The authors, Christopher and Lee, suggest that these risks result from a "lack of supply chain confidence" and that specifically there is a lack of confidence in the following supply chain factors: order cycle time, order current status, demand forecasts, suppliers' capability to deliver, manufacturing capacity, quality of the products, transportation reliability, and the services delivered (Christopher and Lee, 2004). The authors offer the following approaches to reduce risk. Risk can be mitigated by improving information access with greater accuracy and greater visibility throughout the supply chain. The basics of statistical process control can also be used to identify "out-of-control" conditions in the supply chain and to provide alerts (Christopher and Lee, 2004). Contingency plans and corrective actions can be provided to supply chain partners to achieve a more responsive, adaptive supply chain (Christopher and Lee, 2004).

Mason-Jones and Towill (1998) offer a generic model of supply chain uncertainty which is divided into four segments: supply side, manufacturing process, demand side, and control systems (Mason and Towill, 1998). Clearly, the authors have included elements in this model which are consistent with a very broad view of risk and a high-level view of the supply chain.

In a very different approach to risk, Finch (2004) investigated the size of supply chain partners as a factor that may increase risk for the buying firm in the relationship. His findings confirm the need for performing risk assessments and the need to exercise caution when selecting supply chain partners (Finch, 2004).

The literature also offers some excellent examples of companies and their approach to supply chain risks as shown in Table 1.1.

Table 1.1 Company examples for addressing risk

Company	Risk Issue	Method to Address Risk	Author(s)
Benetton	Visibility and control	EDI network	Christopher and Lee 2004
Adaptec	Market risks and loss of market share	Internet technology	Christopher and Lee 2004
Sainsbury (UK)	Access to point-of-sale (POS) data	Extranet	Christopher and Lee 2004
Nokia	Responding to supplier disruption	Contingency planning and execution team	Lee 2004

IBM conducted a study of supply chain executives with surveys completed by 400 supply chain leaders (Butner et al., 2009). Five major supply chain challenges were identified from the survey results. Based solely on response percentages, the top challenges identified were: supply chain visibility (70% of respondents), risk management (60%), increasing customer demands (56%), cost containment (55%), and globalization (43%) (Butner et al., 2009). When the executives ranked these issues by priority, the rankings change slightly. "Cost containment" moves to the top position followed in order as follows: supply chain visibility, risk management, increasing customer demands, and globalization (Butner et al., 2009). In either case, risk is prominent among the issues that the supply chain executives must address for their companies (Butner et al., 2009). More importantly, the other four priorities are topics that are closely interrelated with SCRM.

1.3 Supply Chain Disruptions

A frequently cited definition of supply chain risk describes it as a "variation in the distribution of possible supply chain outcomes, their likelihood, and their subjective value" (Jttner et al., 2003). This builds upon the earlier description from March and Shapira as being "the distribution of possible outcomes, their likelihood and their value" for any management situation where risk is encountered (March and Shapira, 1987).

The research on SCRM covers a very broad range of topics which includes supply chain disruptions as a subcategory. The definition of SCRM has gained significant support in the literature as being: "the identification of potential sources of risk and implementation of appropriate strategies through a coordinated approach among supply chain members, to reduce supply chain vulnerability" (Jttner et al., 2003; Jttner, 2005; Ponomarov and Holcomb, 2009). Risk of supply chain disruptions is described by Hendricks and Singhal (2005a) as "an indication of a firm's inability to match demand and supply."

Supply chain risk may manifest in the form of supply chain disruptions. A supply chain disruption has been described as having "a certain probability of occurrence and is characterized both by its severity and by its direct and indirect effects. Since the resulting detriment is usually a function of time, supply chain disruptions involve time pressure, implying that decisions for mitigation must be made swiftly. ... Depending on its severity, other terms might be applied, such as, glitch, disturbance, accident, disaster, or crisis" (Wagner and Bode, 2008, p. 310).

The literature on SCRM has discussed a very broad range of supply chains. Among those are consumer product manufacturers, major retail firms and a variety of other companies with similar supply chain environments. A limited number have dealt with process industries and continuous flow processes rather than discrete products. Most notably Kleindorfer and Saad conducted a study of the chemical industry and disruptions due to chemical spills or similar events (Kleindorfer and Saad, 2005).

Another commonly used term is supply chain vulnerability. (Chapman et al., 2002) define supply chain vulnerability as: "an exposure to serious disturbance, arising from risks within the supply chain as well as risks external to the supply-chain." They also suggest that risks result from "lack of visibility, lack of ownership, self-imposed chaos, the misapplication of just-in-time practices and inaccurate forecasts" (Chapman et al., 2002). The authors also assert that "the complexity of today's typical supply-chain networks, ... brings with it higher levels of risk and hence vulnerability" (Chapman et al., 2002).

With these definitions as a backdrop, SCRM has garnered significant attention as a research topic and has continued to attract more attention over the last 10 years. Given the relative newness of this topic area, the research is very diverse with many competing viewpoints. It is also not unusual that the topic has been addressed conceptually more often than empirically.

Christopher and Lee (2004) suggest that the increased "vulnerability of supply chains to disturbance or disruption" has the following contributing factors:

- "External events such as wars, strikes or terrorist attacks," and
- "The impact of changes in business strategies" such as
    - "lean" business practices,
    - increased outsourcing decisions, and
    - initiatives to reduce the supplier base (Christopher and Lee, 2004).

Tang and Tomlin (2008) discuss a variety of management initiatives which contribute to greater supply chain vulnerability including:

- Increased product variety, frequent product introductions, and a greater number of sales channels/markets,
- Supply base reduction, use of online procurement options (i.e., e-markets and online auctions), and
- Outsourcing of several functions including manufacturing, information services, and logistics.

Under normal business conditions, the initiatives can be very successful but in a turbulent environment the risk of disruption becomes a concern. These initiatives have created supply chains that are extended over long distances and are more complex which makes them more susceptible to disruptions (Tang and Tomlin, 2008).

Looking at the empirical research about SCRM, three literature reviews provide a summary of the leading examples. Rao and Goldsby (2009), Sodhi et al. (2011), and Monroe (2012) have identified the empirical research on this topic.

For the purposes of this paper, we will now focus on just those articles that are "quantitative" empirical studies. Drawing from the lists of quantitative (or secondary) empirical studies in the two articles referenced above (Monroe, 2012; Rao and Goldsby, 2009; Sodhi et al., 2011) we have a total of nine articles. There are three articles which appear in both lists, one "secondary" article that is unique in the Rao and Goldsby (2009) list and five articles that appear only in the Sodhi et al. (2011) list. Monroe's research adds another 11 articles (Monroe, 2012) to the list of quantitative empirical studies bringing the total to 20. The articles that we add can be seen in Table 1.2.

From Table 1.3, "financial issues" can be identified as the single topic that has received the most attention from the SCRM research in this sample. The impact on shareholder value (and firm market capitalization) is the main topic addressed as evidenced in several articles by Hendricks and Singhal (2003; 2005a; 2005b; 2008; 2009) and Hendricks et al. (2009). This makes the authors the most frequent contributors to the empirical research about SCRM. Their research relies on secondary data to address several related SCRM phenomena. Among these are the impact on stock price and the direct impact on stockholders that occurs as a result of a supply chain disruption event (Hendricks and Singhal, 2003, 2005b, 2008, 2009). A subsequent topic looks at the length of recovery time for the stock's value after a disruptive event (Hendricks and Singhal, 2005a). Another article looked at the impact of a different type of supply chain event—the announcement of "excess inventory" (Hendricks et al., 2009). The analysis of actual events and the impact on companies and their constituencies is an important contribution to the SCRM literature. This type of research to explore the impact of actual events should be extended into other supply chain/organizational segments where data can be related to SCRM constructs.

Table 1.2   Empirical research about SCRM from literature reviews

Author(s) (Year)	Journal	Rao and Goldsby (2009)	Sodhi et al. (2011)	Monroe (2012)
Finch (2004)	SCM	X		
Hendricks and Singhal (2003)	JOM	X	X	
Hendricks and Singhal (2005a)	POM	X	X	
Hendricks and Singhal (2005b)	MgtSci		X	
Kleindorfer and Saad (2005)	POM	X	X	
Wagner and Bode (2008)	JBL		X	
Braunscheidel and Suresh (2009)	JOM		X	
Jiang et al. (2009)	JOM		X	
Ellis et al. (2010)	JOM		X	
de Koster et al. (2011)	JOM			X
Gray et al. (2011)	JOM			X
Hora et al. (2011)	JOM			X
Rao et al. (2011)	JOM			X
Speier et al. (2011)	JOM			X
Hendricks et al. (2009)	JOM			X
Hendricks and Singhal (2009)	MSOM			X
Hendricks and Singhal (2008)	TQM			X
Schoenherr et al. (2008)	JP&SM			X
Wagner and Bode (2006)	JP&SM			X
Jüttner (2005)	IJLM			X

Note: X indicates the articles identified in the literature reviews by authors listed at top of the column.

As described by Rao and Goldsby (2009), the use of secondary data is a dominant choice within the existing empirical research about SCRM. The use of secondary data can be seen in Finch (2004), Hendricks and Singhal (2003), Hendricks and Singhal (2005a; 2005b), and Kleindorfer and Saad (2005). The entire set of Hendricks and Singhal papers (2003; 2005a; 2005b; 2008; 2009; 2009) rely on secondary data to evaluate the financial impact that results from supply chain disruptions and supply chain glitches.

In the subsequent period since 2009, the importance of risk management has remained at the forefront. Disruptions have occurred with regular frequency including hurricanes, earthquakes, tsunamis, volcano eruptions, and floods among the natural disasters (Westrum, 2006). Table 1.4 briefly summarizes the SCRM empirical research for recent years.

Table 1.3 Notable elements in SCRM research

Author(s) Year	Comments on Elements of SCRM Empirical Study
Braunscheidel and Suresh 2009	Survey of 4000 ISM members in top management positions; 218 usable responses; test of 12 hypothesized path coefficients; SEM model includes main factors influencing Firm's Supply Chain Agility (FSCA); External Integration and External Flexibility were first and second major antecedents of FSCA
Ellis et al. 2010	3196 surveys distributed to purchasing professionals; 223 usable responses; test of nine hypotheses to evaluate relationships of several variables and probability of supply disruption or magnitude of supply disruption; CFA and SEM analyses
Finch 2004	Evaluates secondary published data about Information Systems interruptions caused by "natural disasters, accidents and deliberate acts"; from case studies, concludes that larger firms experience greater risk by networking and having small or medium enterprises as partners
Hendricks and Singhal 2003	519 announcements from WSJ & DJNS from 1989 to 2000; glitches—primarily part shortages; test of five hypotheses
Hendricks and Singhal, 2005a	827 announcements from 1989 to 2000; production or shipping delays; test of give hypotheses
Hendricks and Singhal, 2005b	885 announcements from 1992 to 1999; production or shipping delays; test of four hypotheses related to performance metrics
Hendricks and Singhal 2008	838 announcements from 1989 to 2001; loss of 10% of shareholder value for affected firms
Hendricks and Singhal 2009	276 announcements about excess inventory from 1990 to 2002; test of five hypotheses; 73%–74% of sample firms experience negative stock market reaction of nearly 7%
Hendricks et al. 2009	307 supply chain disruptions from 1987 to 1998; new variables in hypotheses include "operations slack," "business diversification," "geographic diversification," and "level of vertical relatedness"
Jiang et al. 2009	634 usable responses from 3000 surveys distributed to Chinese migrant workers in manufacturing sector; labor risk (job satisfaction), labor turnover, and SC risk
Juttner 2005	Quantitative survey of 1700 Institute of Logistics members; 137 usable responses; follow-up with focus groups; results indicate the risk assessment tools in use by logistics professionals
Kleindorfer and Saad 2005	U.S. chemical industry events; 1945 chemical-release accidents between 1995 and 1999 from the total population of 15,219 facilities; over $1 B for business interruption and other indirect costs

(Continued)

Table 1.3   (Continued)

Author(s) Year	Comments on Elements of SCRM Empirical Study
Schoenherr et al. 2008	U.S. manufacturing case study to empirically derive values for 17 risk factors related to outsourcing to China and Mexico; employs AHP to assess the risk factors
Wagner and Bode 2006	Survey of 4946 top-level executives; 760 usable responses; test of three hypotheses about supply chain vulnerability and demand-side risk, supply-side risk and catastrophic risk
Wagner and Bode 2008	Survey of 4946 top-level executives; 760 usable responses; test of five hypotheses about SC risks and SC performance; support for two hypotheses—one for demand-side risk and one for supply-side risk

Table 1.4   Recent empirical research

Author(s) (Year)	Journal	Topic
Ambulkar et al. (2015)	JOM	Resilience to SC disruptions
Bode and Wagner (2015)	JOM	Drivers of SC complexity and SC disruptions
Chen (2014)	POM	Supply disruptions and production efficiencies
Hendricks and Singhal (2014)	POM	Demand-supply mismatches and firm risk
Tang et al. (2014)	POM	Managing disruptions with supply process reliability
Porteous et al. (2015)	POM	Social and environmental compliance at suppliers
Liu et al. (2016)	POM	Build SC resilience with virtual stockpile pooling

1.4 Probability and Impact

The basic definition of risk can be stated as "Risk = Probability of a Loss times the Impact of that Loss" (Manuj and Mentzer, 2008). This definition has also been transformed into a commonly used description as—"probability and impact" and is commonly presented in the form of a 2×2 matrix. Bhattacharya et al. (2009) use the "low probability-high impact" (LPHI) designation throughout their paper and they focus primarily on events that fit the LPHI quadrant. The organization for their perspective can be depicted in a 2×2 matrix as given in Table 1.5.

In general, the LPHI quadrant has received the largest share of attention by researchers. Examples of authors addressing this quadrant include:

- Bhattacharya et al., (2009)
- Christopher and Lee, (2004)

Table 1.5   Probability and impact matrix

	Low Impact	High Impact
High probability of occurrence	HPLI	HPHI
Low probability of occurrence	LPLI	LPHI

Note: HPHI, high probability-high impact; HPLI, high probability-low impact; LPHI, low probability-high impact; LPLI, low probability-low impact.

Table 1.6   Probability and consequences matrix

	Light Consequences	Severe Consequences
High disruption probability		High vulnerability
Low disruption probability	Low vulnerability	

Table 1.7   Examples for probability and consequences matrix

	Light Consequences	Severe Consequences
High disruption probability	Sheffi and Rice (2005) suggest "single port closure and transportation link disruption" as examples in this quadrant.	Sheffi and Rice (2005) suggest "loss of key supplier, labor unrest, economic recession, and visible quality problems" as examples in this quadrant.
Low disruption probability	Sheffi and Rice (2005) suggest "computer virus, wind damage, flood, and workplace violence" as examples in this quadrant.	Sheffi and Rice (2005) suggest "terrorism, earthquakes, supplier bankruptcy, and blizzard" as examples in this quadrant.

- Kleindorfer and Saad, (2005)
- Sheffi and Rice, (2005)
- Tang, (2006)
- Sheffi and Rice (2005) utilize a similar perspective but use slightly different terminology. Sheffi and Rice (2005) use the following 2×2 matrix to label their view of supply chain vulnerability as shown in Table 1.6 and examples in each of the four quadrants can be seen in Table 1.7:

Kraljic (1983) is notable for utilizing a version of the 2×2 matrix for addressing the entire set of suppliers in the purchasing portfolio. While the context is different, using the risk matrix and the portfolio matrix in combination may be useful. From other research fields, risk has come to be associated only with negative events despite the obvious risk related to decisions and investments that lead to favorable results (Kahnemann, 2011; March and Shapira, 1987).

1.5 Developing Resilience in Supply Management

The overall intent of this chapter is to provide the motivation for resilience as seen in the previous sections followed by a functional example where resilience is needed. In this section, the specific focus will be on resilience in supply chain management including the purchasing function with the main objective being supply continuity. This closely parallels and extends the earlier work by Zsidisin et al. (2005) who addressed business continuity planning from the purchasing and supply management perspective.

From the preceding examples from the SCRM literature, there is a clear need for multiple approaches to address supply chain risk. The literature on supply chain resilience is somewhat limited but growing with each successive year.

One publication stated that "supply chain wide risk management ... is not yet recognised as a key element in business continuity planning" (Jttner et al., 2003). That statement was accurate in 2003 but substantial changes have occurred based on actual events mentioned above. The amount of research and the initiatives by practitioners in many industries have also contributed to the knowledge about SCRM in broader terms. New factors need to be considered as the nature of supply chain resilience continues to evolve.

Risk has been defined in the literature by numerous authors across many different research fields. While supply chain risk was never a research topic for these authors, the work of March and Shapira (1987) provides important insights about risk. March and Shapira (1987) are credited with a definition of risk that is often cited. They actually attribute the definition to "classical decision theory" which states that "risk is most commonly conceived as reflecting variation in the distribution of possible outcomes, their likelihoods, and their subjective values" March and Shapira (1987). Juttner et al. (2003) adapted this definition to describe supply chain risk as "the variation in the distribution of possible supply chain outcomes, their likelihood and subjective value."

Quantifying risk is a topic of interest that rivals the interest in the definition of risk. To quantify risk, a widely used expression is stated as "risk is probability times consequence" Kaplan and Garrick (1981). The same authors suggest that this statement is misleading and that a preferred statement is "risk is probability and consequence" (Kaplan and Garrick, 1981). They further suggest that the multiplicative approach might result in a low probability times high consequence event being equated to a high probability times low consequence event when they are clearly very different situations (Kaplan and Garrick, 1981).

From the overall description of risk, the terms "variation," "distribution," and "probability" invoke the image of a probability distribution curve as a representation of risk. Again, Kaplan and Garrick (1981) offer some valuable advice: "a curve is not a big enough concept either. It takes a whole family of curves to fully communicate the idea of risk" (Kaplan and Garrick, 1981). These observations are primary inputs to subsequent parts of the discussion.

There is wide support for the observation that decision makers tend to automatically regard risk as being associated with negative outcomes. March and Shapira (1987) refer to prior work by Shapira who asked managers if they consider all outcomes, negative outcomes or positive outcomes and 80% indicated that they think only of negative outcomes. The availability bias also contributes to viewing risk as negative outcomes because the most memorable (or available) examples will be negative outcomes (Kahnemann, 2011).

The vast majority of research studies on SCRM have dealt with supply chain disruptions. Discussions of tsunamis, earthquakes, hurricanes, and other similar events fall into the quadrant for low probability and high consequence from the 2×2 matrix. Ultimately, that means the main focus then shifts to the consequences of these rare events. This also might be interpreted as research reinforcing the negative outcomes bias. While resilience is needed in these situations, recovery and return to stability may be very challenging amidst complete devastation. Supply continuity planning and resilience will help to address this type of risk but will also address a much wider range of risks as well. Smaller perturbations caused by some event with moderate or low consequences with any probability also require resilience to return to stability or as near to stability as possible under the circumstances.

In a similar manner to risk research, resilience research spans many different fields and has many different descriptions. A sample of the range of resilience definitions is presented in Table 1.8.

Resilience in the context of supply chain management and risk management is not well-defined (Ponomarov and Holcomb, 2009; Wieland and Wallenburg, 2012). The lack of a clear definition is partially due to the limited studies about achieving resilience which means the concept is not fully understood (Wieland and Wallenburg, 2012). A sample of descriptions of resilience that appear in supply chain management research is presented in Table 1.9.

Table 1.8 Resilience described by other fields

Author(s) Year	Field	Description of Resilience
Holling 1973	Ecology	"Resilience determines the persistence of relationships within a system and is a measure of the ability of these systems to absorb changes of state variables, driving variables, and parameters, and still persist." "Stability, on the other hand, is the ability of a system to return to an equilibrium state after a temporary disturbance."
Walker et al. 2004	Ecology	"Resilience is the capacity of a system to absorb disturbance and reorganize while undergoing change so as to still retain essentially the same function, structure, identity, and feedbacks."
Haimes 2009	Engineering	Cites Haimes, 2006
Haimes 2006	Engineering	"The resilience of a system is a manifestation of the states of the system. Perhaps most critically, it is a vector that is time dependent. Resilience is defined as the ability of the system to withstand a major disruption within acceptable degradation parameters and to recover within an acceptable time and composite costs and risks."
Westrum 2006	Engineering	"Resilience is the result of a system (i) preventing adverse consequences, (ii) minimizing adverse consequences, and (iii) recovering quickly from adverse consequences." (Also cited by Haimes, 2009)

Ponomarov and Holcomb (2009) develop the case for supply chain resilience by building upon the conceptual foundations from the ecological perspective as well as the social, psychological, and economic perspectives of resilience. They do not include the engineering perspective about resilience. For the perspectives that are included, the reader is referred to the article by Ponomarov and Holcomb (2009).

Biedermann et al. (2018) address the theoretical and research perspectives related to supply chain resilience. The authors have undertaken a review of resilience research across a wide range of scientific fields. They identify 56 different theories that have been applied within the broader resilience research between 2003 and 2016.

Main elements of resilience as presented in the two articles mentioned above are listed in Table 1.10.

Table 1.9 Resilience described in supply chain management research

Author(s) Year	Field	Description of Resilience
Christopher and Peck 2004	Supply chain management	"It is the ability of a system to return to its original state or move to a new, more desirable state after being disturbed."
Sheffi and Rice 2005	Supply chain management	"Reducing vulnerability means reducing the likelihood of a disruption and increasing resilience—the ability to bounce back from a disruption. Resilience, in turn, can be achieved by either creating redundancy or increasing flexibility."
Ponomarov and Holcomb 2009	Supply chain management	Resilience is defined as: "The adaptive capability of the supply chain to prepare for unexpected events, respond to disruptions, and recover from them by maintaining continuity of operations at the desired level of connectedness and control over structure and function."
Bakshi and Kleindorfer 2009	Supply chain management	"The adverse impact of a disruption can be mitigated by taking steps to reduce the probability of the disruption occurring, or by resuming normal operations quickly and thereby curbing losses."

Table 1.10 Resilience elements from two articles

Ponomarov and Holcomb (2009)—Selected Aspects Identified in Resilience Research	Biedermann et al. (2018)—Supply Chain Resilience Perspectives
Agility, responsiveness	System approach
Visibility	Logistic, transport, and network management
Flexibility/redundancy	Relational governance
Structure and knowledge	Strategy and organization
Reduction of uncertainty, complexity, reengineering	Operations management
Collaboration	Emergency management
Integration, operational capabilities, transparency	Multidisciplinary
	Supplier selection and management
	Ecology
	Psychology

1.6 The Role of Purchasing

Purchasing occupies a prominent position in supply management and must play a central role in developing resilience. Several articles in the late 1990s and early 2000s address SCRM factors without using the term resilience specifically. Subsequent research has incorporated many of those same factors when developing conceptual models of supply chain resilience. The remainder of this section highlights a selection of major factors associated with resilience in the literature.

1.6.1 Supplier Development

Supplier development is a factor that has been studied by numerous researchers. As an example, Krause et al. (1998) analyzed data from 84 companies and compared supplier development practices for strategic and reactive firms. Strategic firms were more likely to co-locate their engineers at suppliers' facilities for supplier development and made direct investments in supplier training programs when compared to reactive firms (Krause et al., 1998). To better serve the strategic firms, suppliers were more likely to add engineers and other personnel to focus on development, assign dedicated personnel to work on performance improvement, and invest their own money in training to further the strategic firms' development proposal (Krause et al., 1998). These results are consistent with the main objectives for supplier development that can also contribute to the development of resilience.

Krause (1999) developed a structural model to evaluate the antecedents of supplier development. "Supplier commitment, expectation of relationship continuity, and effective buyer-supplier communication" were determined to be statistically significant antecedents ultimately leading to supplier development activities (Krause, 1999). Krause et al. (2000) conducted a study and Modi and Mabert (2007) replicated the study to explore supplier development by evaluating operational knowledge transfer and collaborative communication as factors contributing to supplier performance improvement. The results indicate that supplier evaluation and supplier certification efforts are very important prerequisites before operational knowledge transfer can be undertaken (Modi and Mabert, 2007)].

1.6.2 Redundancy

Sheffi and Rice (2005) advocate for redundancy as one factor that will contribute to the development of resilience. Redundancy with regard to suppliers translates to multiple sources rather than a single source for purchased items. Sheffi and Rice (2005) propose building resilience into the supply chain and offer several specific preparation steps to be taken. Redundancy is the first suggestion and takes the form of safety stock inventory, multiple suppliers as sources for the same items, and operating well below capacity to allow for a capacity cushion (Sheffi and Rice, 2005). But they also offer a caution specifically with regard to safety stock inventory that might undermine a company's just-in-time program. The authors describe Flexibility as another main strategy and discuss five elements where flexibility is needed: supply and procurement, conversion, distribution and customer-facing activities, control systems, and the right culture (Sheffi and Rice, 2005). They feel strongly that the potential loss of customers and loss of market share can easily justify the investment in strategic redundancy and significant efforts to develop flexibility.

Tang (2006) also promotes multiple sources as one element of a comprehensive SCRM plan. This requires purchasing to select, qualify, and manage multiple suppliers. The challenge becomes how to share the business volume across multiple suppliers with sufficient volume to hold their interest and to make it worth their while. Most suppliers will not agree to serve only as an emergency backup supplier if they are not receiving an adequate share of the business on a regular basis. Facing this challenge, purchasing must work toward finding a suitable balance for each set of suppliers for a particular item or product family.

1.6.3 Integration

Lee and Whang (2001) assert that "information technology, and in particular, the Internet, plays a key role in furthering the goals of supply chain integration. While the most visible manifestation of the Internet has been in the emergence of electronic commerce as a new retail channel, it is likely that the Internet will have an even more profound impact on business-to-business interaction, especially in the area of supply chain integration." IT (information technology) integration has also been described as "an enabling mechanism that positively impacts supply chain flexibility and supply chain agility" Swafford et al. (2008). The authors further describe their view of integration as a "domino effect" where IT integration "enables a firm to tap its supply chain flexibility which in turn results in higher supply chain agility and ultimately higher competitive business performance" (Swafford et al., 2008). From their results, the combination of IT integration and flexibility will lead to greater supply chain agility when compared to the firm focused on IT integration alone (Swafford et al., 2008).

Braunscheidel and Suresh (2009) include mid-level variables "internal integration" and "external integration" in their structural model of the firm's supply chain agility. The results indicated that both types of integration did have a positive effect on agility.

E-procurement and supplier relationship management (SRM) are two technology elements that must be integrated successfully for the majority of purchasing transactions. In addition to technology, there are human elements that must also be coordinated or integrated.

1.6.4 Visibility

Chopra and Sodhi (2004) identify lack of supply chain visibility and distorted demand information as risk drivers among an extensive list of risk factors. To address forecast risk, they recommend combatting exaggerated forecasts along with improving visibility over the entire supply chain.

The discussion in the previous section regarding Integration is also related to visibility. But Barratt and Oke (2007) caution that information sharing should not be used interchangeably with visibility. The former is an activity and latter is an outcome of that activity (Barratt and Oke, 2007). Barratt and Oke (2007) also develop a model to investigate visibility and they describe several technology-based enablers of visibility but they also focus on other enablers like the human element provided by someone such as the customer service coordinator. Both formal and informal communications by personnel working with customers were viewed as contributing to visibility (Barratt and Oke, 2007).

Gligor and Holcomb (2012) explore the antecedents of supply chain agility with a model that includes supply chain cooperation, supply chain coordination, and supply chain communication. All three factors can also be interpreted as enablers of visibility in addition to producing positive effects on supply chain agility.

1.6.5 Flexibility

A third variable in the Braunscheidel and Suresh (2009) model is "external flexibility" and it also had a positive influence on the firm's supply chain agility in their analysis. Swafford et al. (2006) evaluated "procurement flexibility," "manufacturing flexibility," and "distribution flexibility" in combination with IT integration as the primary factors contributing to supply chain agility.

Gosling et al. (2010) summarize a sample of the supply chain flexibility literature and then propose two main elements: vendor flexibility and sourcing flexibility. Their three-stage model begins with vendor and sourcing flexibility leading to supply chain flexibility. The third level then expands to: new product flexibility, volume flexibility, mix flexibility, delivery flexibility, and access flexibility (Gosling et al., 2010).

1.6.6 Agility

Christopher (2000) offered "agile" as a strategy that is separate and distinct from "lean". He describes the circumstances where the "agile" strategy is needed as "demand is volatile and the requirement for variety is high" which is an unpredictable environment (Christopher, 2000). Christopher (2000) describes agility as "a business-wide capability that embraces organizational structures, information systems, logistics processes and, in particular, mindsets." He also suggests that a key characteristic is "flexibility" that makes agility possible (Christopher, 2000). Christopher and Towill (2001) propose a three-level model of agile supply chain which consists of nine factors for agile manufacturing and ten factors for agile logistics.

1.6.7 Strategic Sourcing

In the course of strategic sourcing, the purchasing team will be seeking suppliers who have a majority of the characteristics on this list (and many others). In one conceptual model, Christopher and Peck (2004) include supply base strategy, sourcing decisions and criteria, and supplier development as one branch of a model of "Resilient Supply Chains". This is a recognition that sourcing will play an important role in most supply chain initiatives.

Swafford et al. (2006) develop another model to evaluate supply chain agility. From their results, supply chain agility "is directly and positively impacted by the degree of flexibility present in the manufacturing and procurement/sourcing processes of the supply chain" (Swafford et al., 2006).

Strategic sourcing was investigated by (Kocabasoglu and Suresh, 2006) by analyzing data from 140 US manufacturing companies. Their results include several of the factors listed here including "information sharing with key supplier and development of key suppliers" (Kocabasoglu and Suresh, 2006). The work by Krause (1999) regarding supplier improvement and development aligns with these results as well. This again reinforces the tight connections that exist between the different factors under consideration by researchers on this topic.

To consider other factors, one recent article offers an interesting view of supply chain resilience. Jain et al. (2017) develop a structural model of supply chain resilience which consists of the following:

- Adaptive capability
- Collaboration
- Trust
- Sustainability
- Risk and revenue sharing
- Information sharing
- Supply chain structure
- Market sensitiveness
- Supply chain agility
- Supply chain visibility
- Risk management culture
- Minimizing uncertainty
- Technological capability among partners

Their results indicate the hypothesized pairwise relationships are "strongly supported" for seven of the hypotheses and "supported" for the other six hypotheses. This is one of the more elaborate models for supply chain resilience which warrants further consideration.

1.7 Maturity Models

The concept of "maturity model" is introduced here to view resilience in a different manner from the majority of the research. The original maturity model is attributed to the Software Engineering Institute at Carnegie Mellon University dating back to the 1980s (Larson and Gray, 2018). The capability maturity model (CMM) provides a way to evaluate an organization's level of evolution for software project management and the use of best practices for software development projects (Larson and Gray, 2018). Several other maturity models have been developed for environments that share some of the common elements of project management but differ significantly from software development. The Project Management Institute has developed a more generic project maturity model which is applicable to many types of projects (Larson and Gray, 2018).

BearingPoint applied the maturity model concept to business processes and utilized a five-stage model very similar to the CMM (Fisher, 2004). The five states of process maturity for the business process maturity model are: siloed, tactically integrated, process driven, optimized enterprise, and intelligent operating network (Fisher, 2004). Notably, the author explains that moving from one level to the next level is not a simple "linear path" and that many "hurdles" must be overcome to reach the next higher maturity state (Fisher, 2004). Lockamy and McCormack (2004) looked at supply chain maturity from the business process perspective which is very appropriate for supply chain processes. Their model is similar to the CMM and includes the following five levels from low maturity to high maturity: ad hoc, defined, linked, integrated, and extended. Another model proposed by PRTM (2001) contains only four levels of maturity. It also matches the integration of supply chain processes as described in the work by Johnson and Whang (2002). The four different levels of the model make the direct connection between maturity level and integration. A comparison of a few of the maturity models and the PRTM model is presented in Table 1.11.

The following section will conclude with a proposed model that merges the concepts of resilience, the concepts from maturity models and some concepts from the very early publications.

1.8 A Proposed Model for Risk-Integration-Resilience

Guided by the literature the components of the proposed supply chain risk-integration-resilience (RIR) model are assembled here. In particular, the RIR model is informed by the full range of literature from different fields and from different time periods. This is done in an effort to include the best practices and the best ideas from prior substantive research.

Today's business environment is characterized by uncertainty, turbulence, and rapidly changing marketplaces. Firm's must manage their supply chains to deal with ever increasing customer expectations and short product lifecycles. Executives must manage in a responsible manner for sustainability and be accountable to the full spectrum of stakeholders. This provides the context for the proposed model and is shown at the bottom of the diagram.

Building the model begins with an informed perspective about risk. Supply chain risk is frequently defined as "the variation in the distribution of possible supply chain outcomes, their likelihood and subjective value" (Jttner et al., 2003). To represent the risk domain, "a whole family of curves" is needed to fully communicate risk (Kaplan and Garrick, 1981). To depict this perspective, the foundation of the model describes supply chain risk conceptually.

Table 1.11 Summary of different levels for maturity models

Maturity Model	Author(s) Year	Level I	Level II	Level III	Level IV	Level V
PMI for project mgt.	Larson and Gray (2018)	Ad hoc	Formal application	Institutionalization of project mgt.	Management of PM system	Optimization of PM system
Business process MM	Fisher (2004)	Siloed	Tactically integrated	Process driven	Optimized enterprise	Intelligent operating network
BPO-SCM	Lockamy and McCormack (2004)	Ad hoc	Defined	Linked	Integrated	Extended
SCPM3	de Oliveira et al. (2011)	Foundation	Structure	Vision	Integration	Dynamics
SC Practice MM	PRTM (2001)	Functional focus	Internal integration	External integration	Cross-enterprise collaboration	Only 4 levels

Integration is a key factor that contributes to resilience in many ways. Systems integration is frequently discussed by using some version of a maturity model. The maturity model from PRTM (2001) is adopted and modified to depict the main phases of integration in the RIR model. The PRTM model consists of only four levels of integration. For the RIR model, the first level has been changed from "Functional Focus" to "Basic Process Focus" and a fifth level has been included to represent the continued effort and the higher degree of integration that is needed to reach the level of "Enhanced Resilient Supply Chain." The five levels of integration are depicted across the top of the RIR model.

Immediately below the integration levels are four different groupings of initiatives which will be required to progress from the current level to the next. Evaluating any individual company will find the company anchored somewhere along the journey through the integration phases. The type of initiatives to be undertaken depends on the current status of the company's integration efforts. Each company must conduct an honest self-audit to determine status and the next steps. The difference in this model is the explicit inclusion of many initiatives and factors that will lead to greater resilience in the supply chain. The initiatives correspond with many of the resilience factors presented earlier including: flexibility, agility, supplier development, supplier collaboration, supplier improvement, and integration.

"Strategic Sourcing" includes supplier selection, supplier development, supplier relationship management (SRM), and supplier improvement programs among a much longer list of supplier focused initiatives. In the model, Strategic Sourcing occupies a central supporting position for all of the activities working toward supply chain resilience. The idea of a Supplier Portfolio is incorporated based on the Kraljic model (1983). "Resilient Strategies" also occupy a prominent position in the model as the basic principles-guiding decision makers. Among these strategies, "Adaptability" is listed alongside "Flexibility," "Agility," and others. The addition of "Adaptability" is inspired by the discussions of resilience by other fields including ecology and engineering. The full RIR model can be seen in Figure 1.1. Taken as a whole, the RIR model serves as a conceptual framework that organizes a significant set of interrelated concepts for developing resilience in supply management.

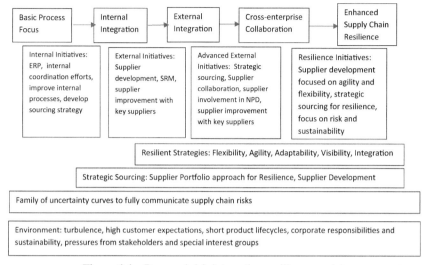

Figure 1.1 Proposed risk-integration-resilience model.

1.9 Conclusion

The background for this chapter relies on SCRM which provides the motivation for developing resilience in supply management. The literature has been reviewed extensively to provide examples of the leading topics in the research. In particular, supply chain disruptions have garnered a substantial share of attention over that last two decades.

To better understand resilience, research from a wide range of disciplines has been consulted to extract definitions and to consider the nuances that are seen through other eyes. As a result, a multifactor, multidisciplinary view of resilience is advocated.

Informed by the other research, a new model of RIR is proposed. This model merges concepts from several different sources and echoes the multifactor, multidisciplinary perspective. Specifically, ideas from ecology, engineering, systems integration, supply chain management, and purchasing management have been blended together in the RIR model.

The topics discussed and the model that is developed are theoretical contributions to the knowledge base. A significant portion of the SCRM literature is organized for easy consumption. The RIR model provides a new, comprehensive view of resilience in the supply chain. In total, the material presented here will be valuable as the foundation for other research.

References

Ambulkar, S., Blackhurst, J. and Grawe, S. (2015). Firm's resilience to supply chain disruptions: Scale development and empirical examination. *Journal of Operations Management*, Vol. 33–34, pp. 111–122.

Bakshi, N. and Kleindorfer, P. (2009). Co-opetition and investment for supply-chain resilience. *Production and Operations Management,* Vol. 18, No. 6, pp. 583–603.

Baramichai, M., Zimmers, E. W. and Marangos, C. A. (2007). Agile supply chain transformation matrix: an integrated tool for creating an agile enterprise. *Supply Chain Management: An International Journal*, Vol. 12, No. 5, 334–348.

Barratt, M. and Oke, A. (2007). Antecedents of supply chain visibility in retail supply chains: A resource-based theory perspective. *Journal of Operations Management*, Vol. 25, pp. 1217–1233.

Bhattacharya, A., Geraghty, J. and Young, P. (2009). On the analytical framework of resilient supply-chain network assessing excursion events, *2009 Third Asia International Conference on Modelling & Simulation*, IEEE Computer Society.

Biedermann, L., Kotzab, H. and Pettit, T. J. (2018). Theory landscape and research perspectives in current supply chain resilience research, *International Conference on Dynamics in Logistics* and appears in *Dynamics in Logistics*, Michael Freitag, Herbert Kotzab and Jurgen Pannek, editors, pp. 26–33.

Bode, C. and Wagner, S. M. (2015). Structural drivers of upstream supply chain complexity and the frequency of supply chain disruptions. *Journal of Operations Management*, Vol. 36, pp. 215–228.

Braunscheidel, M. J. and Suresh, N. C. (2009). The Organizational Antecedents of a Firm's Supply Chain Agility for Risk Mitigation and Response. *Journal of Operations Management*. Vol. 27, pp. 119–140.

Butner, K. F. R., Casey, A., Sundaram, K., Kinser, Ch., Meyer, B. et al. (2009). The Smarter Supply Chain of the Future: Global Chief Supply Chain Officer Study. IBM Corporation, IBM Global Services.

Chapman, P., Christopher, M., Jüttner, U., Peck, H. and Wilding, R. (2002). Identifying and managing supply-chain vulnerability. *Logistics and Transport Focus*, Vol. 4, No. 4, p. 59.

Chen, Y.-J. (2014). Supply disruptions heterogeneous beliefs, and production efficiencies. *Production and Operations Management*, Vol. 23, No. 1, pp. 127–137.

Chopra, S. and Sodhi, M. S. (2004). Managing risk to avoid supply-chain breakdown. *Sloan Management Review*, Fall, pp. 53–61.

Christopher, M. (2000). The agile supply chain: Competing in volatile markets. *Industrial Marketing Management*, Vol. 29, No. 1, pp. 37–44.

Christopher, M. and Lee, H. (2004). Mitigating supply chain risk through improved confidence. *International Journal of Physical Distribution & Logistics Management*, Vol. 34, No. 5, pp. 388–396.

Christopher, M. and Peck, H. (2004). Building the resilient supply chain. *International Journal of Logistics Management*, Vol. 15, No. 2, pp. 1–13.

Christopher, M. and Towill, D. (2001). An integrated model for the design of agile supply chains. *International Journal of Physical Distribution & Logistics Management*, Vol. 31, Issue 4, pp. 235–246.

Craighead, C. W., Blackhurst, J., Rungtusanatham, M. J. and Handfield, R. B. (2007). The severity of supply chain disruptions: Design characteristics and mitigation capabilities. *Decision Sciences*, Vol. 38, No. 1, pp. 131–156.

de Koster, R., Stam, D. and Balk, B. M. (2011). Accidents happen: The influence of safety specific transformational leadership, safety consciousness, and hazard reducing systems on warehouse accidents. *Journal of Operations Management,* Vol. 29, pp. 753–765.

de Oliveira, M. P. V., Ladeira, M. B. and McCormack, K. P. (2011). The Supply Chain Process Management Maturity Model—SCPM3, *Supply Chain Management*, DOI: 10.5772/18961. Available from: https://www.intech open.com/books/supply-chain-management-pathways-for-research-and-pr actice/the-supply-chain-process-management-maturity-model-scpm3

Ellis, S. C., Henry, R. M. and Shockley, J., (2010). Buyer perceptions of supply disruption risk: A behavioral view and empirical assessment. *Journal of Operations Management*, Vol. 28, pp. 34–46.

Finch, P. (2004). Supply chain risk management. *Supply Chain Management: An International Journal*, Vol. 9, No. 2, pp. 183–196.

Fisher, D. M. (2004). The business process maturity model. A practical approach for identifying opportunities for optimization. BPTrends [online)], http://www.bptrends.com/resources_publications.cfm. (Accessed 12 May 2018).

Gligor, D. M. and Holcomb, M. C. (2012). Antecedents and consequences of supply chain agility: Establishing the link to firm performance. *Journal of Business Logistics*, Vol. 33, No. 4, pp. 295–308.

Gosling, J., Purvis, L. and Naim, M. M. (2010). Supply chain flexibility as a determinant of supplier selection. *International Journal of Production Economics*, Vol. 128, No. 1, pp. 11–21.

Gray, J. V., Roth, A. V. and Leiblein, M. J. (2011). Quality risk in offshore manufacturing: Evidence from the pharmaceutical industry. *Journal of Operations Management*, Vol. 29, pp. 737–752.

Haimes, Y. Y. (2006). On the definition of vulnerabilities in measuring risks to infrastructures. *Risk Analysis*, Vol. 26, No. 2, pp. 293–296.

Haimes, Y. Y. (2009). On the definition of resilience in systems. *Risk Analysis*, Vol. 29, No. 4, pp. 498–501.

Hendricks, K. B. and Singhal, V. R. (2003). The effect of supply chain glitches on shareholder wealth. *Journal of Operations Management*, Vol. 21, No. 5, pp. 501–522.

Hendricks, K. B. and Singhal, V. R. (2005a). An empirical analysis of the effect of supply chain disruptions on long-run stock price performance and equity risk of the firm. *Production and Operations Management*, Vol. 14, No. 1, pp. 35–52.

Hendricks, K. B. and Singhal, V. R. (2005b), Association between supply chain glitches and operating performance. *Management Science*, Vol. 51, No. 5, pp. 695–711.

Hendricks, K. B. and V. R. Singhal. (2008). The effect of supply chain disruptions on shareholder value. *Total Quality Management*, Vol. 19, No. 7–8, pp. 777–791.

Hendricks, K. B. and Singhal, V. R. (2009). Demand-Supply mismatches and stock market reaction: Evidence from excess inventory announcements. *Manufacturing & Service Operations Management*, Vol. 11, No. 3, pp. 509–524.

Hendricks, K. B., Singhal, V. R. and Zhang, R. (2009). The effect of operational slack, diversification, and vertical relatedness on the stock market reaction to supply chain disruptions. *Journal of Operations Management*, Vol. 27, No. 3, pp. 233–246.

Hendricks, K. B. and Singhal, V. R. (2014). The effect of demand-supply mismatches on firm risk. *Production and Operations Management*, Vol. 23, No. 12, pp. 2137–2151.

Holling, C. S. (1973), Resilience and stability of ecological systems, *Annual Review of Ecology and Systematics*, Vol. 4, pp. 1–23.

Hora, M., Bapuji, H. and Roth, A. V. (2011). Safety hazard and time to recall: The role of recall strategy, product defect type, and supply chain player in the U.S. toy industry. *Journal of Operations Management*, Vol. 29, pp. 766–777.

Jain, V., Kumar, S., Soni, U. and Chandra, C. (2017). Supply chain resilience: model development and empirical analysis. *International Journal of Production Research*, Vol. 55, No. 22, pp. 6779–6800.

Jiang, B., Baker, R. C. and Frazier, G. V. (2009). An analysis of job dissatisfaction and turnover to reduce global supply chain risk: Evidence from China. *Journal of Operations Management*, Vol. 27, No. 2, pp. 169–184.

Johnson, M. E. and Whang, S. (2002). E-Business and supply chain management: An overview and framework. *Production and Operations Management*, Vol. 11, No. 4, pp. 413–423.

Jüttner, U. (2005). Supply chain risk management: Understanding the business requirements from a practitioner perspective. *International Journal of Logistics Management*, 16, 1, 120–141.

Jüttner, U., Peck, H. and Christopher, M. (2003). Supply chain risk management: Outlining an agenda for future research. *International Journal of Logistics: Research & Applications*, Vol. 6, No. 4, pp. 197–210.

Kahnemann, D. (2011). *Thinking, Fast and Slow*. Farrar, Straus and Giroux: New York.

Kaplan, S. and Garrick, B. J. (1981). On the quantitative definition of risk. *Risk Analysis*, Vol. 1, No. 1, pp. 11–27.

Kleindorfer, P. R. and Saad, G. H. (2005). Managing disruption risks in supply chains. *Production and Operations Management*, Vol. 14, No. 1, pp. 53–68.

Kocabasoglu, C. and Suresh, N. C. (2006). Strategic sourcing: An empirical investigation of the concept and its practices in U.S. manufacturing firms. *Journal of Supply Chain Management*, Vol. 42, No. 2, pp. 4–16.

Kraljic, P. (1983). Purchasing must become supply management. *Harvard Business Review*, Vol. 61, No. 5, pp. 109–117.

Krause, D. R. (1999). The antecedents of buying firms' efforts to improve suppliers. *Journal of Operations Management*, Vol. 17, pp. 205–224.

Krause, D. R., Handfield, R. B. and Scannell, T. V. (1998). An empirical investigation of supplier development: Reactive and strategic processes. *Journal of Operations Management*, Vol. 17, pp. 39–58.

Krause, D. R., Scannell, T. V. and Calantone, R. J. (2000). A structural analysis of the effectiveness of buying firms' strategies to improve supplier performance. *Decision Sciences*, Vol. 31, No. 1, pp. 33–55.

Lambert, D. M., Cooper, M. C. and Pugh, J. D. (1998). Supply chain management: Implementation issues and research opportunities. *International Journal of Logistics Management*, Vol. 9, No. 2, p. 1.

Larson, E. W. and Gray, C. F. (2018). *Project Management: The Managerial Process*, 7th edition, McGraw-Hill/Irwin: New York.

Lee, H. L. (2004). The triple-A supply chain. *Harvard Business Review*.

Lee, H. L. and Billington, C. (1993). Material management in decentralized supply chains. *Operations Research*, Vol. 41, pp. 835–847.

Lee, H. L. and Whang, S. (2001). E-Business and Supply Chain Integration, *Stanford Global Supply Chain Management Forum*, SGSCMF-W2–2001.

Liu, F., Song, J.-S. and Tong, J. D. (2016). Building supply chain resilience through virtual stockpile pooling. *Production and Operations Management*, Vol. 25, No. 10, pp. 1745–1762.

Lockamy, A. and McCormack, K. (2004). The development of a supply chain management process maturity model using the concepts of business process orientation. *Supply Chain Management: An International Journal*, Vol. 9, pp. 272–278.

Manuj, I. and Mentzer, J. T. (2008). Global supply chain risk management strategies. *International Journal of Physical Distribution & Logistics Management*, Vol. 38, No. 3, pp. 192–223.

March, J. G. and Shapira, Z. (1987). Managerial perspectives on risk and risk taking. *Management Science*, Vol. 33, No. 11, pp. 1404–1418.

Mason-Jones, R. and Towill, D. R. (1998). Shrinking the supply chain uncertainty circle. *Control*, pp. 7–22.

Modi, S. B. and Mabert, V. A. (2007). Supplier development: Improving supplier performance through knowledge transfer. *Journal of Operations Management*, Vol. 25, pp. 42–64.

Monroe, R. W. (2012). Supply chain risk management: A review of the empirical research, *POMS 23rd Annual Conference*, Chicago, Illinois, USA, April 23.

Pereira, C. R., Christopher, M. and da Silva, A. L. (2014). Achieving supply chain resilience: The role of procurement. *Supply Chain Management: An International Journal*, Vol. 19, No. 5/6, pp. 626–642.

Ponomarov, S. Y. and Holcomb, M. C. (2009). Understanding the concept of supply chain resilience. *The International Journal of Logistics Management*, Vol. 20, No. 1, pp. 124–143.

Porteus, A. H., Rammohan, S. V. and Lee, H. L. (2015). Carrots or sticks? Improving social and environmental compliance at suppliers through incentives and penalties. *Production and Operations Management*, Vol. 24, No. 9, pp. 1402–1413.

PRTM (Pittiglio, Rabin, Todd and McGrath). (2001). The Performance Measurement Group, LLC, by Bob Moncrieff, Director, PRTM and

Mark Stonich, Principal, PRTM, Supply-Chain Practice Maturity Model and Performance Assessment, *Supply Chain Council Webcast*, November 6, 2001. [online]. ftp://www.inf.fh-dortmund.de/pub/contributors/schlichtherle/Literatur/SCM/6903-MV_SCCWEBCAST.pdf (accessed 12 May 2018).

Rao, S. and T. J. Goldsby. (2009). Supply chain risks: A review and typology. *International Journal of Logistics Management*, Vol. 20, No. 1, pp. 97–123.

Rao, S., Griffis, S. E. and Goldsby, T. J. (2011). Failure to deliver? Linking online order fulfillment glitches with future purchase behavior. *Journal of Operations Management*, Vol. 29, pp. 692–703.

Schoenherr, T., Tummala, V. M. R. and Harrison, T. P. (2008). Assessing supply chain risks with the analytic hierarchy process: Providing decision support for the offshoring decision by a US manufacturing company. *Journal of Purchasing & Supply Management*, Vol. 14, pp. 100–111.

Sheffi, Y. and Rice, J. (2005). A supply chain view of the resilient enterprise. *Sloan Management Review*, Vol. 47, No. 1, pp. 41–48.

Sodhi, M. M. S., Son, B.-G. and Tang, C. S. (2011). Researchers perspectives on supply chain risk management. *Production and Operations Management*. Vol. 21, No. 4, pp. 1–13.

Speier, C., Whipple, J. M., Closs, D. J. and Voss, M. D. (2011). Global supply chain design considerations: Mitigating product safety and security risks. *Journal of Operations Management*. Vol. 29, pp. 721–736.

Swafford, P. M., Ghosh, S. and Murthy, N. (2006). The antecedents of supply chain agility of a firm: Scale development and model testing. *Journal of Operations Management*, Vol. 24, No. 2, pp. 170–188.

Swafford, P. M., Ghosh, S. and Murthy, N. (2008). Achieving supply chain agility through IT integration and flexibility. *International Journal of Production Economics,* Vol. 116, No. 2, pp. 288–297.

Tang, C. S. (2006). Perspectives in supply chain risk management. *International Journal of Production Economics*, Vol. 103, No. 2, pp. 451–488.

Tang, C. S. and Tomlin, B. (2008). The power of flexibility for mitigating supply chain risks. *International Journal of Production Economics*, Vol. 116, Issue 1, pp. 12–27.

Tang, S. Y., Gurnani, H. and Gupta, D. (2014). Managing disruptions in decentralized supply chains with endogenous supply process reliability. *Production and Operations Management*, Vol. 23, No. 7, pp. 1198–1211.

Wagner, S. M. and Bode, C. (2006). An empirical investigation into supply chain vulnerability. *Journal of Purchasing & Supply Management*, Vol. 12, pp. 301–312.

Wagner, S. M. and Bode, C. (2008). An empirical examination of supply chain performance along several dimensions of risk. *Journal of Business Logistics*. Vol. 29, No. 1, pp. 307–325.

Walker, B., Holling, C. S., Carpenter, S. R. and Kinzig, A. (2004). Resilience, adaptability and transformability in social-ecological systems. *Ecology and Society*, Vol. 9, No. 2, pp. 5–16.

Westrum, R. (2006). A typology of resilience situations. In Hollnagel E, Woods DD, Leveson N (eds). *Resilience Engineering: Concepts and Precepts*. Aldershot, UK: Ashgate Press, pp. 49–60.

Wieland, A. and Wallenburg, C. M. (2012). Dealing with supply chain risks: Linking risk management practices and strategies to performance. *International Journal of Physical Distribution & Logistics Management*, Vol. 42, No. 10, pp. 887–905.

Zsidisin, G. A., Melnyk, S. A. and Ragatz, G. L. (2005). An institutional theory perspective of business continuity planning for purchasing and supply management. *International Journal of Production Research*, Vol. 43, No. 16, pp. 3401–3420.

2

Designing a Resilient Consumer Product System

Kanchan Das

College of Engineering and Technology, East Carolina University,
NC 27858, USA
E-mail: DASK@ecu.edu

This research plans to include a model-based system design framework to contain the failures and disruptions faced by the consumer product supply chains (SCs) and improve their resilience by examining the way they may fail. The containment measures for developing system resilience considers options to improve absorption capability, adaptability, and recoverability for each of the system functions for the SC considering its dependence and interconnectedness with other systems. The research investigates literature to study past failures and disruptions faced by such systems. In addition, the research follows examples of strategies that successful SCs pursued or have been pursuing to withstand or overcome such disruption events. The research plans strategy to mitigate the disruption effects and include recovery options when disruptions cannot be contained by absorption capabilities. In order to fulfill customer requirements by making the system resilient and monitor status of SC performance before, during, and after recovery from the disruptions, the research formulated resilience performance coefficients for the SC functions. The model connected the resilience coefficient performance indices for the SC functions with the planning process variables for the functions. A numerical example illustrates applicability of the model.

2.1 Introduction

Based on definition of Holling (1973) resilience "is a measure of the persistence of systems and of their ability to absorb change and disturbance

31

and still maintain the same relationships between populations or state variable" in the ecological perspectives. This research proposes system resilience definitions "as the ability of a system to absorb effects of disruptions and plan strategy for adaptation to the disruptions process to reduce its effects on system performance and quick recovery to reach a better or previous stage when absorption is not possible."

Modern society needs continuous flow of essential goods, food, automobile, smooth running and maintenance of information and healthcare systems for its continuation and survival (Nan and Sansavini, 2017). But these critical systems are exposed to several natural and manmade disruptions (Chopra and Sodhi, 2014; Li et al., 2017; Linkov et al., 2017; Sheffi and Rice, 2005). As such it is crucial that supply chains (SCs) of these essential items are needed to be resilient for the welfare and well-being of the society. Critical systems, such as consumer product system, water supply system, transportation, telecommunication, health care system, and information system that provide essential services to our society have become more interdependent, complex due to globalization and to fulfill increasing social needs (Nan and Sansavini, 2017). As the systems are increasingly more complex, their disruptions are more damaging in terms of severity to the society as covered in highly publicized events including 2003 Black out of North America, 2004 Indian Ocean Tsunami, 2010 Haiti and Chile earthquakes, 2010 eruption of Icelandic Volcano, 2012 Hurricane Sandy (Bhamra et al., 2011; Raj et al., 2015). In recent years, disruptions are also becoming more unpredictable, frequent and thus more damaging (Shen and Tang, 2015). As such, it is imperative that critical systems such as consumer product should be resilient to ensure continuation of services to our society. Due to increasing complexity, interdependencies, frequency, and severity of disruptions, in addition to varying nature of system functions, an innovative resilient system design model-based approach should be there to support supply of essential items to society.

Past research mostly emphasized risk management for a system to make it resilient, see for example, research of Chopra and Sodhi (2004) and literature review of Bhamra et al. (2011). To combat the disasters such as devastated effects of Hurricane Katrina and Rita and recover from their effects US Government (federal and State) have been pursuing risk management approach (Raj et al., 2015). Such approach depends on probability-based approach considering past data to develop risk management. There are several uncertainties in connecting data to develop or estimate probability, especially in the face of several interdependencies and interconnectedness of modern SCs. Paradigm of resilience is considered as a better approach than risk management to enable the systems or part of the systems to withstand disruption risk and recover quickly Raj et al. (2015). In this research, we follow Resilience paradigm-based approach to create resilient systems. The National Academy of Sciences (NAS) (2012), in its manuscript *Disaster Resilience: A National Imperative*, defines resilience as "the ability to plan and prepare for, absorb, recover from and adapt to adverse events." In addition, in resilience creation our objective is to deal mostly with the unknown disasters and preparing a system for shocks that could challenge its ability to perform efficiently under stress. So, it is apparent that in a system resilient development design research each function of the system should be built to have absorptive, adaptive, and recovery capability for containing potential or unknown disasters. Based on above functional systems should be planned such that it will be prepared for absorption, adaptation, and recovery.

This research contributed by including such preparation steps as included in the definition of resilience by NAS while planning each functional (supply, manufacturing, and distribution) management systems. For an example, in the supply management planning it included supplier flexibility, supply location flexibility, and not including a supplier from calamity-prone location to develop absorptive capability. It includes other factors and planning steps for adaptability and recovery capability to contain disruptions that cannot be contained by absorptive steps. The remaining part of the chapter includes the following: the next section includes study of background literature. After study of literature it address methodology to be followed for designing a resilient SC for consumer product system. A numerical example illustrates applicability of the resilience system design approach after methodology, and conclusions and discussion on the chapter is presented in the last section.

2.2 Study of Background Literature

Systems design with resilience may be considered a new research area since emphasis on resilience creation got increased attention from academia and government organizations during the last decade. But number of research in this area has grown to an extent that it may be treated as one of the very rich research subject area. The two streams of research that created background of this research are: Resilience system design approach for SC management and resilience performance indices for the SC.

2.2.1 Resilience System Design Approach

Resilience system design approach is mostly applied in systems engineering design areas. A resilient engineered system is considered to have ability to sense, quickly respond, and withstand adverse events, which is the goal of engineering design (Mackenzie and Hu, 2018). In recent years such resilience system design approach is taken in ecology (Webb, 2007), organizational capability building by including flexibility to quickly recover from disruptions by the SC as covered in Sheffi and Rice (2005). In the research of Sheffi and Rice (2005), recommendations based on disruption profile phases for preparation, disruption event, first response, and the impacts are elaborated based on resilient system design approach for an enterprise. The approaches included are also in line with resilience approach recommended by The National Academy Science (2012) for disaster resilience. Systems resilience approach may be expanded to connect with firm's profit through modeling-based approach proposed in (Mackenzie and Hu, 2018).

Resilience systems design approach may be expanded to enterprise information system (Zhang and Lin, 2010). The unique feature of Zhang and Lin's research in (Zhang and Lin, 2010) is the addition of five principles, that are also included in the Sheffi and Rice (2005) and in Sheffi's research in reference (Sheffi, 2015). These principles include having certain degree of redundancy, control and management of redundancy, sensor for monitoring and controlling, inclusion of predicting capability, and actuator for implementing control. These principles can be adapted to resilient SC system design.

Systems engineering-based resilient system design are mostly mathematical modeling based as may be observed in Nan and Sansavini (2017); Youn (2011). Engineering systems are often complex, interdependent, and interconnected which is subjected to cascading effects when it is hit by a catastrophic event. In addition, it is an emergent system that is dynamically evolving (Park et al., 2013) to address societal needs in critical systems like power supply, food supply management, and health care system. Extant research appreciates these facts considering interdependencies, changing boundary conditions that increase the severity of low-frequency high-impact hazards and proposes to integrate resilience in engineering system by following a recursive process that includes sensing, anticipation, learning, and adaptation considering systematic adaptation of the system not incremental and gradual (Leveson, 2004). Given the today's increasing complexity for interconnected engineering systems in addition to increasing frequency and severity of catastrophes sensing may be considered very limiting to decide to go for creating resilience systems. Monitoring the system for changing risk profile as recommended in (Madani and Jackson, 2009) should be the way, where monitoring system includes updated knowledge base for increased system complexity and changed risk profile for disaster by including inputs from news media, social media for system's inability to perform. As such engineering systems should be designed by integrating resilience system creation approach. Literature finding also indicate that the drivers for systems resilience may also be qualitative, such as staff cooperation and collaboration during disruptive events, level of preparation against natural disaster, in addition to quantitative as discussed in Hussaini et al. (2016) using Bayesian approach to quantify system resilience by connecting qualitative and quantitative assessments.

We may mention here that our model-based research includes most of the factors and considerations included in the above resilience systems research. We have not considered qualitative factors presented in Hussaini et al. (2016). We emphasized quantitative factors such that SC managers may take a target-based resilience development plan.

2.2.2 Resilient Performance Indices for the Systems

Operational disruptions impact SCs ability to match supply with demand. SC resilience is created to enable it to contain disruptions and continue to fulfill customer requirements. So, it is important to measure and monitor resilience performance of a SC to understand its status with respect to fulfilling customer requirements. In fact, performance metrics may be considered as the bottom line which is verifiable measure for the status of the firm for operational performance feedback (Hanson et al., 2011; Munoz and Dunbar, 2015). Munoz and Dunbar (2015) studied a quantification of operational SC resilience in a multidimensional multitier SC situation using a simulation-based approach. The study used a hypothetical discrete manufacturing SC. In their simulations they assumed: manufacturer and retailer of the considered SC may be either fully operational or not at all, and customer satisfies their demand from other sources when retailer is not functional which mimicked fill rate affected by SC failure. The simulation model inflicted disruptions for nonfunctional situations. Disruption duration assumed to be three periods (periods equivalent to cycle time). End of disruption assumed as operation restoration and start of recovery. SC performance data collected as the order fill rates at the customer and retailer tiers by the calculation of a first-order exponential smoothing equation with a 0.95 weight percentage. Performance for recovery is measured as the time to return to acceptable performance, say time t1–t2 after disruption ends. Impact is measured as the difference of performance level at t1 and t2. The performance level at t1 means starting of exponential graph, t2 level means highest depth of the graph/chart for disruption; performance loss is the area of graph. They ran simulation for 200 runs. Finally, they used a structural equation to compute average SC resilience performance by combining performance, performance loss, and impact and weight factors. As is apparent the procedure used in Munoz and Dunbar's research is complex and not practicable. Soni et al. (2014) used a graph theory-based model that considered enabler of resilience and their interrelationship for analysis in an interpretive structural modeling (ISM) approach. The model quantified resilience in a single numerical index. For identifying enablers and common relationship among them Soni et al.'s research conducted a survey among experts from Indian firms and academia. Based on survey responses for 10 enablers, such as (1) agility, (2) collaboration among players, (3) information sharing, (4) sustainability in SC, (5) risk and revenue sharing, (6) trust among players, (7) SC visibility, (8) risk management culture, (9) adaptive capability, (10) SC structure were selected and used in their model considering SC resilience index (SCRI) = f(resilience enablers). The directed graph for SC considered nodes $N = N_i$'s as resilience enablers and edges N_{ij}'s as the dependence of enablers. The enablers listed above are represented as $N1, N2. \ldots$ For analyzing enablers in structural self-interaction matrix four symbols V: enabler i to help to achieve j; A: enabler j to achieve i; X: i,j' s will help achieve each other; O: i,j's are unrelated. In the next step, resilience variable characteristics matrix (VCM-RM) were developed based on effect of various enablers. Later they considered off diagonal matrix of A with N_{ij} and B with diagonal element N_i; VCM-RM as $C = B$ - A. Then they considered resilience variable permanent matrix: VPM-RM, $N = B + A$ and permanent representation $N^* = A + B + C + D + E$, where $D = \sum_{1,2.....10} (N12, N23), (N34, N43), (N5, N6...)$; and $E = \sum_{1,2.....10}$ sum of $(N12, N23, N34, N51 + N15...)$. Then they quantified N_i and N_{ij}'s. The SCRI = Per N^*.

Using the proposed approach in Munoz and Dunbar (2015) resilience index of two SCs may be compared. As may be observed that the SCRI may provide good managerial insights but the process of determination is too complex to be used practically.

Yilmaz-Borekci et al. (2015) studied supplier resilience in the context of supplier buyer relationship through an empirical study. The research collected data from 183 suppliers from Turkey firms. The study used collected data on three resilience construct dimensions: structural resilience, mostly organization capability of supplier for cash flow, product development, maintenance as the next dimension for organizational capability for product realization or processing; third dimension is like the first one related to resources factor. For exploring validation of dimensions for supplier resilience, the research conducted exploratory factor analysis to establish their dimensions such that the study provides some clear sense to SC mangers. Simchi Levi et al. (2014) developed a risk exposure index and applied it at Ford Motor Company. It evaluates an organization's vulnerability to certain disruptions. In our model, we consider various disruption reasons and integrate absorptive, adaptive, and recovery provisions as applicable, whereas Simchi-Levi et al. (2014) evaluated vulnerability irrespective of cause(s). One more fundamental difference is that our research evaluates resilience performance based on factors (for example: flexibility for resilience creation) such that an SC can strategically decide how much resilience to include for its targeted resilience performance. Their model is not applicable to overall SC resiliency performance index evaluation, as it basically evaluates time to recover (TTR) when a vulnerability occurs. In a sense, their approach is not a planning model, but rather an after-affect analysis approach. Using a TTR-based model, an organization can determine TTR and decide to provide related importance to a vulnerability based on TTR value. This decision for imparting importance can then be used for future planning.

A failure mode and effects analysis (FMEA)-based risk measurement and prioritization approach for managing probable and foreseeable risks was proposed in Bradley (2014). Although Bradley's research Bradley (2014) did not cover resilience measures, it is reviewed here because it measured/assessed risk to contain SC disruptions. Bradley (2014) considered the probability of risk based on expert opinions, and the risk effect based on potential revenue loss to mitigate the risk consequences.

Our resilience performance index % of market requirement compliance will provide clearer idea to any SC manager on their resilience performance status compared to extant research. For any SC, objective of resilience creation is to successfully satisfy market requirements. Since resilience performance index value according to our model is the cumulative effect of absorptive, adaptive, and recovery strategy-based steps, and each resilience steps for any function may also be monitored and investigated individually, SC managers can initiate target-based steps for improving resilience performance.

2.3 Methodology

The research investigates literature to find out failures and disruptions faced by consumer product-based SCs. It studies literature that includes academic publications, newspaper reports, reports from creditable international organizations. UNISDR (2013) and publicly available documents from FEMA, NOAA, and Government Departments (NIAC, 2009; NIPP, 2009; NOAA, 2017) that deal with critical systems and infrastructures to identify disruptions and their effects on system performance over the past years. In our research, effects on system resilience performance is estimated based on % of customer requirements served before and after the disruptions. Based on such objectives and findings from past research it formulates a model-based design framework for a resilient SC.

For creating resilience to potential disruptions an organization should build absorptive, adaptive, and recovery capability (Chopra and Sodhi, 2014, 2004; Fiksel et al., 2015; Nan and Sansavini, 2017; Sheffi and Rice, 2005). Absorptive capability includes creation of flexibility, redundancy, product quality, and reliability assurance system, keeping inventory for withstanding, and weathering away disruption effects (Fiksel et al., 2015; Nan and Sansavini, 2017; Sheffi and Rice, 2005). Adaptive capability includes strategy, actions or early readiness to minimize consequences of a disruption effects which could not be inhibited/weathered away by absorptive steps. Such strategy includes installing a proactive emergency action plan; creating capability to modify/divert operations in the event of a disruption where possible, segmenting or decentralizing operations to take support from not affected segments or area or outsourced vendor; detecting potential for disruptions and creating contingency plan (Chopra and Sodhi, 2014; Fiksel et al., 2015; Sheffi, 2015). Next is the recovery that includes restorative capability or ability of a system to be repaired to take it back to original or a better stage, if possible, as quickly as possible; real time monitoring, establishing control tower like expert hub for supervisory control and data acquisition to quickly react for repairing or initiating recovery steps by taking external help or using alternative facility to bypass part of the disruptions; and or capability to go for lower damage stage for resources by quick restoration. Decentralization, segmentation of the system, crisis management, and business continuity plan can also enhance recovery (Chopra and Sodhi, 2014; Fiksel et al., 2015; Sheffi, 2015). This research formulated the model-based resilience plan by including above-discussed three stages of disruption management recommended in the literature for supply, production, and distribution management. By normal functioning of a system it is meant that the system's coefficient of performance is at the expected level. As such for following up of a system from absorption to adaptability, or recovery of a system from a disruption event; coefficient of performance will guide a company to initiate required steps to bring the system to its normal function.

2.3.1 Problem Statement

Let P be the set of consumer product manufactured by a SC using a set of inputs I supplied by a set of suppliers in a set of plants J. The product is marketed in a set of M markets through a set of distribution centers (DCs) K to retail outlets R. Let CMD_{pmr} be the overall demand for product $p \in P$ from the retailer $r \in R$ of market $m \in M$. The DCs are spread over the entire country to have optimum distribution time. We assume that DCs are in pre-decided location and this study does not address location of DCs. The DCs include select number DCs in safe locations also. Retail outlets are considered customer. System performance includes % fulfillment of product demand in the market. As such objective of the research is to maximize coefficient of performance of the overall system which is the combination of coefficient performances for the functional systems. The research objective also includes optimization of total cost for the SC of the consumer product system and conducting an analysis to understand how total cost of SC and resilience creation influence each other. Since resilience creation involves investment that has long- and short-term implications, such cost vs resilience performance analysis will provide managerial insights for deciding and building resilience in an organization. This Resilient Consumer Product Systems Planning Model first formulates equations that create basis for planning of overall business systems. It then formulates planning of the functional systems for the supply, manufacturing, and distribution of product with their coefficient of performance. *Notations* presented in Table 2.1 are used in defining the model equations and constraints.

2.3.2 The Resilient Consumer Product Systems Planning Model

Equations (2.1) and (2.2) create basis for the overall system planning. Equation (2.1) plans fulfilling of customer demand in various markets by sending products from standard DCs and the safe DCs that are located in safe zones with respect to most of the natural calamities. Here y_{pkmr} and $y_{pk'mr}$ denote the product distributed from standard DC (k) and DC (k') at safe locations (mentioned as safe DC) to the retailer in the markets.

$$\sum_{k \in K} y_{pkmr} + \sum_{k \in K'} y_{pk'mr} = CMD_{pmr} \forall p, m, r \qquad (2.1)$$

Table 2.1   *Notations* used in the model

Indices

$b \in B$: business continuity plan

$c \in C$: calamities

$d \in D$: outsourced vendors for supplying product

$e \in E$: emergency and planned recovery steps

$h \in H$: network partner plants

$i \in I$: inputs/components

$j \in J$: manufacturing plants operated by the SC to manufacture SC product

$k \in K$: distribution centers (DCs) for warehousing products and distributing them from there to retailers; $k' \in K'$: DCs in safe locations

$l \in L$: locations for suppliers

$m \in M$: markets

$n \in N$: quality attributes tracked for evaluating supplier quality performance

$p \in P$: product

$q \in Q$: attributes for evaluating and monitoring manufacturing plant's processing capability for quality, reliability, safety, and similar metrics

$r \in R$: sales outlets, retailers

$s \in S$: enlisted suppliers; $s' \in S'$: standby suppliers

$t \in T$: scenarios

Parameters and variables

acSR: acceptable level of supply system resilience coefficient of performance (*SSCP*)

$a_s = 1$, if supplier 1 is quality affiliated based on desired level of performance; 0 otherwise

acMSCP: acceptable level of manufacturing system resilience *MSCP*

$ax_{is'}$ auxiliary 0/1 variable for input i to decide requirements supply from standby supplier s'

BN, BNN: big numbers used in constraints

$bz_{is'} = 1$, if standby supplier s' is assigned supply order for input i; 0 otherwise

CAE_{pdm}: capacity of outsourced vendor d of market m to supply product p

$CALH_{jc}$: a 0/1 parameter for inputting 1 by user or some automatic knowledge base if calamity c affects plant j, 0 otherwise

CAW_{pk}: capacity of DC k to accommodate product p

CDR_{pkr}: cost of distributing product p from DC k to sales outlet r

c_{is}: cost supplying input i by supplier s.

CLK_{kct}: % calamity effect on distributed product from DC k at scenario t

CM_{pj}: cost of manufacturing product p in plant j

CMD_{pmr}: demand of product p from customer/retailer at market m

$COLB_p$: average cost of product from network partner

col_{pht}: % of production capacity for product p that may be obtained at scenario t from collaborative network partner plant h

$colx_{pj}$ product p available from collaborative network partner to be sent to SC plant j

---

(Continued)

Table 2.1 (*Continued*)

con_{bs}: 1, if business continuity step b is installed and applied for supplier s; 0 otherwise

CPL_p: average cost of product from pooling plant capacities

DA_s: desired level of overall performance by a suppliers s

DIS_{lc}: average per year frequency of calamity c for location l based on the past data from state or government sources.

$DISPL_{pmr}$: product p to be sent to a starved retailer r of market m by pooling capacities of partially and nonaffected DCs; starving due to non/ partial supply of product by calamity affected DCs

dp_l: 1, if a location l had frequency of disruption cause or calamity events during last 10 years exceeded set limits and became calamity-prone location; 0 otherwise.

EFF_{ct}: effect of calamity c for scenario t

emr_{pjmr}: emergency supply of product p from plant j to retailer r in market m when the DC allocated becomes calamity affected

$ext_{pj} = 1$, if product p is sent from plant j under emergency plan

ez_{is} is effective amount of the input i received by the SC after ordering z_{is} to supplier s

FCE_{ie}: fixed cost emergency services e is installed for input i

FM_{pj}: cost of setting up production of product p in plant j

COV_p: average cost of product p from outsourced vendors

CTR_{pjk}: cost of transporting product p from plant j to DC k

DCA_{kc}: a 0/1 parameter for inputting 1by user or some automatic knowledgebase if calamity c affects DC k, 0 otherwise

$DISC_{ci}$: a 0/1 parameter for inputting 1by user or some automatic knowledgebase if calamity c really affected input supply i, 0 otherwise

DL_{lc}: maximum set limit for frequency of calamity c for a location l to be defined as calamity prone

dw_k: 1, if standard DC k is open; 0 otherwise

$em_{ei} = 1$, if emergency/recovery option e is implemented to handle failure situation for i; 0 otherwise

ex_{pjk} effective amount of product p manufactured in SC plant j to transport to standard DC k after considering effect of disruptions and containment measures

ey_{pkmr}: effective amount of product p distributed to retail outlet r in market m from DC k after considering effect of disruptions and containment measures

FCB_{sb}: Fixed cost for installing emergency procedure b for supplier

FEM_{pdm} fixed cost for ordering product p to outsourced vendor d of market m

FS_{is}: cost of ordering input to supplier s

(*Continued*)

Table 2.1 (*Continued*)

Indices

FSB_{is}': fixed cost of ordering input i to standby supplier s'

FWR_{kmr}: fixed cost for establishing contract with retailer r of market m to buy product from SC to be supplied from DC k

LC_{sl}: location identifier parameter, becomes 1 if supplier s is from location 1, 0 otherwise

$pcol_t$: probability of scenario t for obtaining support from network partner plants

qa_{ns}: performance of the supplier s for the quality attributes n

QU: qualifying minimum score set by the SC for each quality attribute a plant must comply with for fulfill customer quality expectation

sbz_{is}: input i supply by standby supplier s'

spz_{ie}: emergency support for input i to be taken for the emergency and recovery step e

$sdw_{k'}$: 1, if Safe DC k' is open; 0 otherwise

$u_{is} = 1$, if supplier s is assigned supply order for input i, 0 otherwise

x_{pjk} product p manufactured in SC plant j to transport to standard DC k

$xpool_p$ product p to be taken from pooling capacities of plants

$y_{pk'mr}$: product p distributed to retailer r in market m from safe DC k

z_{is}: input i ordered to supplier

$v_{pj.} = 1$, if production of product p is assigned to plant j, 0 otherwise

$vn_{pdm} = 1$, if outsourced vendor d of market m is allocated product p, 0 otherwise

FW_k: fixed cost for opening DC k for operation

$g_{ie} = 1$, if emergency and recovery step e is initiated for input i; 0 otherwise

h: holding cost for inventory per unit

MX_{ie}: maximum possible support for input i from emergency and recovery step e

$p_{t.}$: probability for scenario t

QSR_{qj}: plant performance for each attribute q in a 0 to 1 scale; q includes: scrap rate; plant reliability in terms of OEE; process capability; inventory turnover, and such others

RE_{isb}: estimated contribution of business continuity plan b for supplier s related to input i

SS_i: safety stock for input i to maintain a set service level

$sdrw_{k'mr}$: 1, if safe DC k' is allocated to retailer r of market m; 0 otherwise

$sy_{pk'mr}$: product p distributed from safe DC k' to retailer r in market m

μz_i, $(\sigma z_i)^2$: assumed mean usage and variance following normal distribution for input i

$xs_{pjk'}$ product p manufactured in SC plant j to transport to SDC k'

y_{pkmr}: product p distributed to retailer r in market m from DC k

up_j: 1, if plant j qualified to be the capable plant for complying with SC's quality standard, 0 otherwise

xd_{pj} product p available from the decentralized/outsourced vendors to be sent to plant j

Equation (2.2) balances product supplied from manufacturing plant to standard and safe DCs with the product distributed from the DCs to retailers in the markets for fulfilling customer demand.

$$\sum_{j\in J}(x_{pjk} + xs_{pjk'}) = \sum_{m\in M}\sum_{r\in R}(y_{pkmr} + y_{pk'mr}) \quad \forall p, k, k' \qquad (2.2)$$

2.3.2.1 Resilient supply management systems planning

Please see definitions of variables and parameters in Table 2.1 for following model equations. Supply management approach first determines the inputs requirements in Equation (2.3) and assigns input supply orders in constraint (2.4).

$$\sum_{p\in P}\rho_{pi}\sum_{j\in J}\sum_{k\in k}(x_{pjk} + xs_{pjk'}) = \sum_{s\in S} z_{is} \quad \forall i \qquad (2.3)$$

$$z_{is} \leq u_{is}CAS_{is} \quad \forall i, s \qquad (2.4)$$

It then takes resilient system planning-based approach to include absorption capability by creating flexibility in terms of including more than one suppliers in constraint (2.5) and suppliers from more than one location in constraints (2.6) and (2.6.a)

$$\sum_{s\in S} u_{is} \geq 1 \quad \forall i \qquad (2.5)$$

$$u_{is}LC_{sl} = ucs_{isl} \quad \forall i, s, l \qquad (2.6)$$

$$\sum_{s\in S}\sum_{l\in L} ucs_{isl} \geq 1 \quad \forall i \qquad (2.6.a)$$

Constraint (2.7) identifies calamity-prone location. Constraint (2.8) ensures assignment of a supplier to non-calamity-prone location for a supply item. It may be mentioned here that by not selecting supplier from calamity-prone location but selecting supplier from more than one location will also support in adaptability and recovery from supply system failure due to a severe hurricane in some supplier's location. Supporting supplier in the installation of emergency system and business continuity plan for recovery capability of the supplier are also included in the planning model (to be followed).

$$dp_l \leq LC_{sl} \quad \forall s, l \qquad (2.7)$$

$$\sum_{c\in C} DIS_{lc}u_{is}LC_{sl} \leq DL_l dp_l \quad \forall i, s, l \qquad (2.8)$$

The supply management approach also establishes supplier selection procedure considering critical to quality and critical to business metrics following Das (2011) such that supply items are quality ensured. Such system will make the system resilient to quality and safety system failure that may create media attention. SCs for consumer product system should pursue such supplier selection and affiliation procedure in creating partnering relation with high-quality supplier, that will support them in creating absorption and developing adaptation capability of the supply management system for any emergency created from a sudden situation, like failure of some supply consignment due to natural calamity, some transportation system failure or hurricane. The partners will support in such situations. Let the SC tracks quality records for the past supplies by the suppliers for timely delivery, supply quantity discrepancies, supplier's plant performance, environmental performances, and supplier's business continuity plan to affiliate the quality capable suppliers and assign orders for procuring inputs. Let $n \in N$ are the set of quality attributes tracked by SC by defining suitable performance evaluation criteria and finally transforming them in a 0–1 scale by normalizing them. Since these attributes interacts with each other: let qa_{ns} denotes the performance of the supplier s for the attributes n. Let DA_s be the desired level of overall performance by a suppliers s, $a_s = 1$, if supplier s is affiliated or found suitable to assign supply orders based on desired level of performance; 0 otherwise. Constraint (2.9) identifies affiliated suppliers suitable for assigning supply orders based on the discussed procedure. Constraint (2.10) assigns supply orders only to affiliated suppliers.

$$DA_s a_s \leq \prod_{n \in N} qa_{ns} \quad \forall s \qquad (2.9)$$

$$u_{is} \leq a_s \quad \forall i, s \qquad (2.10)$$

Constraint (2.11) estimates input to be ordered to supplier considering on-hand inventory of input in SC stock and safety stock to be maintained. Safety stock used in constraint (2.11) is estimated assuming of usage of input i to follow Normal distribution $i \sim N(\mu z_i, (\sigma z_i)^2)$ with mean usage μz_i and variance $(\sigma z_i)^2$. SC orders z_{is} quantity of input i to supplier s and adjusts with on-hand inventory zo_i to maintain a safety stock (SS_i) with a defined % stock out; for an example for 2% stock out: $z_{0.98} \sigma z_i \sqrt{LT}$ defines safety stock with supplier lead time LT.

$$z_{is} + SS_i - zo_i \leq u_{is}BN \quad \forall i, s \tag{2.11}$$

Constraint (2.12) estimates emergency support as obtained from emergency support sources at the adaptability and recovery phase of disaster impact on supply management. Adaptive capability of the system is the self-organization to minimize disruption impacts. Such supply system adaptability may be improved by keeping inventory (such as SS_i in constraint (2.11)). An adaptive step slows the failure and related effect, thus getting time for taking other control measures for quicker recovery steps. Constraint (2.13) ensures applying one emergency measure for each input. This is planned just to avoid creation of inventory and expenses from that.

$$spz_{ei} \leq g_{ei}MX_{ei} \quad \forall e, i \tag{2.12}$$

$$\sum_{e \in E} g_{ei} \leq 1 \quad \forall i \tag{2.13}$$

Other emergency support defined in constraint (2.12) includes: (i) informing marketing team and production department to go slow due to a potential supply system failure from one or more suppliers for some disruption effects; (ii) buying inputs from alternative supplier to run everything smoothly within a short interval of time for the applicable inputs; (iii) sending an emergency technical team to support disruption affected supplier for quick recovery; (iv) going for spot buying; (v) assuming SC includes decentralized production system as a part of recovery plan and sister organization factories created by such plan are the other options. In addition, the SC (buyer here) support suppliers to include set of lean and green-based steps, business continuity plan, information network in their plants for further improving supply management resilience. Such supplier support items include: (a) installing business continuity plan for each supplier organization; (b) providing training and supporting suppliers in implementing lean and environment friendly sustainable practices for production, waste disposal process, their input procurement from their suppliers (tier 2); (c) empowering their employees; (d) helping the suppliers to create decentralized resources system; (e) including the suppliers within the information network of SC to monitor supply quality, supplier performance, supply status, predict calamity initiate control steps for recovery by taking outside help. Support from these business continuity plans are estimated in the last term of Equation (2.14) if there input supply is affected by calamities. Constraint (2.15) ensures applying one business continuity plan for a supplier.

$$ez_{is} = z_{is} \left(1 - \sum_{t \in T} \sum_{c \in C} DISC_{ci} eff_{ct} pt \right) + \sum_{e \in E} spz_{ei}$$

$$+ \sum_{c \in C} DISC_{ci} \sum_{b \in B} con_{sb} RE_{isb} \quad \forall\, i, s \qquad (2.14)$$

$$\sum_{b \in B} con_{bs} \leq 1 \quad \forall\, s \qquad (2.15)$$

Constraints (2.16) and (2.17) decide if input i supply sbz_{is} from standby supplier s is needed or not for keeping the supply system coefficient $SSCP$ defined in Equation (2.19) at a desired level through an auxiliary 0/1 variable (ax_{is}). Constraint (2.18) assigns supply orders to standby supplier within the limitations of its capacity confirms assigned input.

$$ez_{is} + sbz_{is'} - z_{is} \leq MN(1 - ax_{is'}) \quad \forall\, i, s \qquad (2.16)$$

$$sbz_{is'} \leq ax_{is'} MN \quad \forall\, i, s' \qquad (2.17)$$

$$sbz_{is'} \leq bz_{is'} BN \quad \forall\, i, s' \qquad (2.18)$$

Supply System Coefficient of Performance (SSCP): Supply management system planning effect for including absorption, adaptability, and recovery capability are considered in defining ($SSCP$) in Equation (2.19). Supply system resilience is a precondition for the SC resilience and overall performance of the SC. $SSCP$ will provide option to evaluate and monitor supply system for its resilience performance.

$$SSCP = \frac{\sum_{i \in I} \left(\sum_{s \in S} (ez_{is} + sbz_{is}) \right)}{\sum_{p \in P} \sum_{i \in I} \rho_{pi} \sum_{m \in M} \sum_{r \in R} CMD_{pmr}} \qquad (2.19)$$

Numerator for $SSCP$ in Equation (2.19) includes effective supply amount of input from suppliers based on constraints (2.11), (2.14), (2.16), and supply from standby supplier based on constraints (2.10) and (2.11). Denominator estimates total inputs needed for fulfilling demand. Further to above, for creating resilient supply management system SCs (buying organizations) should support suppliers through resources and providing required training for installing effective emergency handling procedure and business continuity plan.

2.3.2.2 Resilient manufacturing system planning model

Manufacturing system realizes product by allocating production to plants through constraint (2.20).

$$\sum_{k \in K} x_{pjk} + \sum_{k' \in K'} xs_{pjk'} \leq v_{pj} BNN \quad \forall p, j \qquad (2.20)$$

Constraint (2.21) balances product transported from plants to standard and safe DCs with the product distributed to retailers in markets for ensuring fulfillment of market demand by the product realized in the manufacturing plants.

$$\sum_{j \in J}(x_{pjk} + xs_{pjk'}) = \sum_{m \in M} \sum_{r \in R}(y_{pkmr} + y_{pk'mr}) \quad \forall p, k, k' \qquad (2.21)$$

Realization of product with desired quality level for fulfilling market demand is the primary objective of the manufacturing systems planning function. Manufacturing system resilience includes planning absorptive capability to withstand and/or weather away potential disruptions, such as: natural calamity, terrorist attack, internal sabotages, quality failures, and transportation failure. Following the definition of resilience when the disruptions cannot be contained by the planned absorptive capabilities, SC needs to plan adaptive and recovery strategy to make the system resilient such that disruption effects are minimized and the system can recover and continue its performance or reach a better stage by effective recovery. The model-based approach proposed in this study includes the strategy to create the resilient manufacturing system by including absorptive, adaptive, and recovery capability such that manufacturing system coefficient of performance (*MSCP*) remains at an acceptable level. Constraint (2.22) ensures allocation of more than one plant for manufacturing a product for creating capacity flexibility as a part of absorptive capacity.

$$\sum_{j \in J} v_{pj} \geq 1 \quad \forall p \qquad (2.22)$$

Constraint (2.23) defines the effective quantity of product obtainable from manufacturing system after considering effects of calamities and supports from planned absorptive, adaptive, and recovery steps when such steps are required. The supports included in the equation are: pooled capacity from plants ($xpool_p$), supply of product from outsourced vendor (xd_{pj}) supports from network partners ($colx_{pj}$). Each of these supports is subject to occurrence and nonoccurrence of disaster effect, as defined in the equation by including $CALH_{jc}$. In a live system, a user or an intelligent knowledge base integrated input system will input 0 or 1 for $CALH_{jc}$ when a calamity happened or emerging. As is apparent from Equation (2.23) that these supports are planned to take effect when needed. Such that it does not create extra financial burden for the system for its future continuity of functioning.

$$
\begin{aligned}
\sum_{k \in K} ex_{pjk} \\
= \sum_{k \in K} (x_{pjk} + xs_{pjk'}) * \left(1 - \sum_{c \in C} CALH_{jc} \sum_{s \in S} EF_{cs} P_s \right) \qquad \forall j, c \\
+ xpool_p * \sum_{c \in C} (CAL_{jc} + (xd_{pj} + CAP_{pj} colx_{pj}) * CAL_{jc})
\end{aligned}
$$

$$(2.23)$$

Constraint (2.24) limits $xpool_p$ by the spare able percentage of overall pooled capacity from the plants.

$$
xpool_p \leq spare_p * \sum_{j' \in J'} CAP_{pj'} \quad \forall p \qquad (2.24)
$$

Constraint (2.25) limits product to be obtained from outsourced vendors based on their capacity.

$$
\sum_{j \in J} xd_{pj} \leq \sum_{d \in D} \sum_{m \in M} vn_{pdm} CAE_{pdm} \quad \forall p \qquad (2.25)
$$

Constraint (2.26) ensures one vendor from each market to supply product in each market to adapt to the situation when central plants and pooled plants are disrupted.

$$
\sum_{e \in E} vn_{pdm} = 1 \quad \forall p, m \qquad (2.26)
$$

Constraint (2.27) estimates adaptive product support obtainable from collaborative network partner plants. Col_{pht} is the % of production capacity of product p support obtainable from network partner h at various scenario t

$$
\sum_{j \in J} colx_{pj} = \sum_{h \in H} \sum_{t \in T} col_{pht} pcol_t \quad \forall p \qquad (2.27)
$$

We may mention here that constraints (2.22)–(2.24) provide absorptive capability to manufacturing. Constraint (2.23) also supports adaptive and recovery steps. Support from external vendor (xd_{pj}) and network partner plants ($colx_{pj}$) will be obtainable just like the decentralized and distant resources when production in the central plants and redundant capacities around the plants fail or needed to be stopped due to disaster impact. Manufacturing system of a SC needs to include and enhance adaptability and recoverability to have resilience. Manufacturing adaptability and recoverability can be improved by including early detection ability. Early detection supports a company to get time to prepare and to get external support to mitigate the disruptions within a shorter interval. Such detection ability may be obtained by including potential business scanning software that will consider inputs from TV, radio news, social media on natural calamity and various disruption events. Adaptation and recovery capability may be obtained by decentralizing its manufacturing facility and distribution centers locations. Such decentralization may also include planning separate manufacturing plants for each segment. In decentralized and segmented manufacturing plan, unaffected plants at different location can support manufacturing plants in affected locations. The decentralization and segmentation is a strategic level decision to be planned for a long-time horizon. The product support from decentralized outsourced vendor defined in constraints (2.23), (2.25), and (2.26) are planned to provide adaptive support effectively. Similar support may be very much obtained from network partner plants as defined in constraints (2.23) and (2.29). Product support as defined in constraint (2.24) from pooled plants, outsourced vendor, network partner plants takes effect in the case parameter CAL_{jc} used as the multiplier in constraint (2.24) becomes 1, meaning natural calamity or some other disruptions could damage normal resources, and supports are taken at that point in the absorption as well as adaptive phase. For enhancement of recovery capacity, literature (Fiksel, 2015) recommended to develop collaborating relationship with other suitable organizations for mutual benefit during a disaster instance like natural calamity, terrorist attack or employee noncooperation. Constraint (2.27) estimated amount of production capacity $colx_{pj}$ available from following similar approach, which contributes in resilience creation through balancing constraint (2.24). CAL_{jc} in constraint (2.24) may be used for early detection such that support xd_{pj} and $colx_{pj}$ may be quickly organized for quick adaptability and recovery steps.

The manufacturing management system planning also includes quality-based plant capability determination procedure considering critical to quality and critical to business metrics following [32] such that products from manufacturing system are quality ensured. Constraint (2.28) identifies quality capable plants considering discussed critical to quality and critical to business metrics based on lean, green, quality, reliability, and safety-based criteria. Similar to supplier affiliation, plant capability determination also considers interaction of the metrics in the constraint (2.28). Please refer to description of QSR_{qj} in *Notations* to have ideas on the metrics considered.

$$QUup_j \leq \prod_{q \in Q} QSR_{qj} \quad \forall j \tag{2.28}$$

Constraint (2.29) ensures allocation of production to a quality capable plant only.

$$v_{pj} \leq up_j \quad \forall p, j \tag{2.29}$$

Manufacturing Systems Coefficient of Performance (MSCP): Manufacturing system resilience is the crucial component for overall system resilience for a SC. Any disruption in manufacturing system creates ripple effect on entire SC operation. Equation (2.30) formulates coefficient of performance for the manufacturing systems (*MSCP*) that will provide option to evaluate and monitor manufacturing system for its resilience performance.

$$MSCP = \frac{\sum\limits_{p \in P} \sum\limits_{j \in J} \sum\limits_{k \in K} ex_{pjk}}{\sum\limits_{p \in P} \sum\limits_{m \in M} \sum\limits_{r \in R} CMD_{pmr}} + (acSR - SSCP) \tag{2.30}$$

Denominator for the *MSCP* is the overall product demand to fulfill market requirements. Numerator for *MSCP* includes: effective amount of product transported to standard and safe DCs after considering calamity effects and the containment measures as defined in Equation (2.23); second term is the shortage of product due to lower value of supply management resilient performance (*SSCP*) compared to acceptable level (*acSR*). ex_{pjk} included in the numerator for *MSCP* defined in Equation (2.23) connects absorptive, adaptive, and recovery steps planned in constraints (2.22)–(2.29) including product support $xpool_p$, xd_{pj}, and $colx_{pj}$ no other variables are needed in the numerator . Explanations on the second term: since resilience performance of supply system will negatively affect manufacturing quantity and overall fulfillment of demand this term is included. Let SC defines *acSR* as the acceptable level of *SSCP*. So when $acSR > SSCP$, $(acSR - SSCP)D_p$ estimates negative effect of supply systems, which will be translated to effect of input supply shortage on manufactured product. Such shortage will affect overall market demand. So, for higher *MSCP*, SC should invest in supplier development and improving their performance to make $acSR - SSCP$ nearer to zero.

2.3.2.3 Resilient product distribution system planning

Distribution of product to fulfill customer demand has been planned in Equation (2.1). Balancing product to be distributed to customer with the product received from manufacturing plants has been defined in Equation (2.2). Transportation from plants to standard and safe DCs is planned in constraints (2.31) and (2.32). According to these constraints product is transported within the capacity limitation of DCs and only when the DCs are open.

$$\sum_{j \in J} x_{pjk} \leq dw_k CAW_{pk} \quad \forall p, k \tag{2.31}$$

$$\sum_{j \in J} xs_{pjk'} \leq sdw_{k'} CAW_{pk} \quad \forall p, j \tag{2.32}$$

In this model, the SC distributes products to retail outlets $r \in R$ in each market $m \in M$. SC distribution systems are exposed to several risks and disruptions including: transportation system failure due to severe weather conditions; border congestion; custom employee strike; huge accident which may block the distribution route or disconnects the transportation route due to flood or other calamity; closure of DC due to natural calamity. Constraints (2.33) and (2.34) plan allocation of standard and safe DCs to supply retail outlets.

$$\sum_{p \in P} y_{pkmr} \leq w_{kmr} BN2 \quad \forall k, r \tag{2.33}$$

$$\sum_{p \in P} sy_{pk'mr} \leq sw_{k'mr} BN3 \quad \forall k', r \tag{2.34}$$

For improving absorptive capability of the distribution system, the model creates distribution flexibility by assigning more than one standard DC to a sales outlet by constraint (2.35). Since safe DCs are assumed to not have any calamity impact assigning a safe location DC by constraint (2.36) to each sales outlet will improve such absorption capability.

$$\sum_{k \in K} w_{kmr} \geq 1 \quad \forall\, m, r \tag{2.35}$$

$$\sum_{k' \in K'} sw_{k'mr} = 1 \quad \forall\, m, r \tag{2.36}$$

Constraints (2.37) and (2.38) ensure assigning only the open DCs to retail outlets.

$$w_{kmr} \leq dw_k \quad \forall\, k, m, r \tag{2.37}$$
$$sw_{k'mr} \leq sdw_{k'} \quad \forall\, k', m, r \tag{2.38}$$

When a DC is affected by a calamity or non-supply of product due to one of the reasons (transportation disruption by heavy snow fall, obstruction of product flow due border congestions, heavy traffic jam, road blockade due a big accident) for non-supply of product, the model plans extra stock to be distributed for the affected sales outlets by pooling less affected and nonaffected DCs. It may be mentioned here that safe DCs are assumed to be nonaffected. The safe DCs allocated to a sales outlet any way provides some support to retail outlets starving for product due to non-supply of product from the standard DCs allocated for its product supply. Such supply will act as the available safety stock kept in safe DCs. Let CLA_{ks} is theparameter in terms (% affect in the range 0–1 at scenario s) for the affected DCs when DCA_{kc} becomes 1 for occurrence of a calamity, probability of such scenario is p_s. This 0/1 value for DCA_{kc} is inputted by a person or a system responsible for monitoring. Constraint (2.39) estimates pooled product to be supplied to starved sales outlets due to a calamity-based disruptions of distribution from DCs allocated to the outlets. Constraint (2.40) ensures selection of correct DC-outlet combinations.

$$\sum_{k \in K} \left(1 - \sum_{c \in C} DCA_{kc} \sum_{s \in s} CLK_{kcs} p_s \right) CAW_{pk} = displ_{pmr} \quad \forall\, p, m, r \tag{2.39}$$

$$displ_{pmr} \leq w_{kmr} BN3 \quad \forall\, p, k, m, r \tag{2.40}$$

After pooling product from partially affected and nonaffected DCs, the organization may plan to send product directly from manufacturing plant to retailer in the markets as an emergency measure if still the retailer is unable to comply with market requirements. Let emr_{pjmr} be the product to be sent from the plant to retailer. Constraint (2.41) defines the emergency supply possible from plant to retail outlets based on its capacity limitations.

$$\sum_{m\in M}\sum_{r\in R} emr_{pjmr} \leq ext_{pj}CAP_{pj} \quad \forall p,j \tag{2.41}$$

Constraint (2.42) estimates the effective amount of product distributed from standard DCs to retailer r in market m considering calamity effects at various scenario and select containment measures (last two terms of the equation). Constraints (2.43) and (2.44) plan emergency supply from plants to retail outlets if other containment measures unable to fulfill demand from retailers.

$$\sum_{k\in K} ey_{pkmr} = \sum_{k\in K} y_{pkmr}\left(1 - \sum_{c\in C} DCA_{kc}\sum_{s\in S} CLK_{kcs}P_s\right)$$
$$+ displ_{pmr} + \sum_{j\in J} emr_{pjmr} \quad \forall p,m,r \tag{2.42}$$

$$-\sum_{k} ey_{pkmr} + displ_{pmr} + emr_{pjmr} \leq BNN(1 - bx_{pmr}) \quad \forall p,m,r$$
$$\tag{2.43}$$

$$\sum_{j\in J} emr_{pjmr} \leq BNNNbx_{pmr} \quad \forall p,m,r \tag{2.44}$$

Coefficient of Performance for Distribution System Resilience: Coefficient of performance for distribution system resilience, $DSCP$ may be defined by Equation (2.45).

$$DSCP = \frac{\sum_{p\in P}\sum_{r\in R}\left(\sum_{k\in K} ey_{pkmr} + \sum_{k'\in K'} sy_{pk'mr}\right)}{\sum_{p\in P}\sum_{m\in M}\sum_{r\in R} CMD_{pmr}} + (acMSCP - MSCP)$$
$$\tag{2.45}$$

Denominator for $DSCP$ is the overall product demand in each of the system component since objective of system resilience design is to fulfill customer demand and thus improve overall system performance. The first component in the numerator is the effective amount of product distributed from standard DCs based on Equations (2.42)–(2.44) to sales outlets, the second item is product distributed from safe DCs to retailers. The last item is the shortage effect due to resiliency gap of manufacturing system which will reduce product to be obtained from manufacturing to distribute to sales outlets. It may be mentioned here that distribution resiliency performance index Equation (2.45) considers calamity effects and containment measures for distribution system by ey_{pjkmr} as defined in constraints (2.42)–(2.44) in a combined way.

Objective Functions of the Model: The study includes two objectives, Objective function I maximizing overall resilience as defined in Equation (2.46) and Objective function 2 minimizing overall SC cost for its optimum operation and creating resilience as defined in Equation (2.48).

Objective Function 1: Maximize Overall Resilience
Coefficient of Performance (*OVCP*)   (2.46)

Overall resilience coefficient of performance (*OVCP*) is defined in terms of its components in Equation (2.47).

$$OVCP = SSCP + MSCP + DSCP \qquad (2.47)$$

where: *SSCP* is the supply system coefficient of performance for resilience as defined in Equation (2.19); *MSCP* is the manufacturing system coefficient of performance for resilience as defined in Equation (2.30); and *DSCP* is the distribution system coefficient of performance for resilience as defined in Equation (2.45).

Objective Function 2: Minimize Total Cost for Resilience
Integrated Supply Chain (*TRISC*)   (2.48)

Total cost for resilience integrated supply chain (*TRISC*) is defined in Equation (2.49) in terms of its components

$$TRISC = TRISUP + TRIMFG + TRIDIST \qquad (2.49)$$

where: *TRISUP* is the resilience integrated supply system management cost as defined in Equation (2.50); *TRIMFG* is the resilience integrated manufacturing system management cost as defined in Equation (2.51); and *TRIDIST* is the resilience integrated distribution system management cost as defined in Equation (2.52).

TRISUP, the first component of *TRISC* as defined in Equation (2.50) computes resilience integrated supply management cost by considering cost of buying effective quantity of inputs from the assigned suppliers, fixed cost for assigning the input orders to suppliers, cost of carrying safety stock for the inputs, fixed cost for establishing business continuity plan for supplier's operation, and cost for installing and operationalize emergency procedure for managing input shortage related issues.

$$
\begin{aligned}
TRISUP = \sum_{i \in I} \Bigg(&\sum_{s \in S} (ez_{is}c_{is} + FS_{is}u_{is}) \\
&+ \sum_{s' \in S'} (c_{is'}sbz_{is'} + bz_{is'}FSB_{is'}) + SS_i * h \Bigg) \\
&+ \sum_{s \in S} \sum_{b \in B} con_{sb}FCB_{sb} \\
&+ \sum_{e \in E} \sum_{i \in I} (FCE_{ie}em_{ie} + CEMS_{ie}spz_{ie}) \qquad (2.50)
\end{aligned}
$$

TRIMFG, the second component of *TRISC* defined in Equation (2.51) computes resilience integrated manufacturing system cost by considering costs for manufacturing product to be transported to standard DCs as well as to safe DCs, fixed cost setting a up the plant for production, cost of procuring product from outsourced vendor, cost of procuring product from collaborative partners, and cost of realizing product by pooling the capacity of plants operated by the SC.

$$
\begin{aligned}
TRIMFG = \sum_{p \in P} \sum_{j \in J} \Bigg(&CM_{pj} \sum_{k \in K} x_{pjk} + \sum_{k' \in K'} xs_{pjk'} \Bigg) \\
&+ \sum_{p \in P} \sum_{j \in J} FM_{pj}vpp_{pj} + \sum_{p \in P} COV_p \sum_{j \in J} xd_{pj} \\
&+ COLB_p \sum_{j \in J} colx_{pj} + \sum_{j \in J} CPl_p xpool_p \qquad (2.51)
\end{aligned}
$$

The third component of *TRISC* defined in Equation (2.52) computes *TRIDIST* (resilience-integrated transportation and distribution cost), by considering costs for transporting product from SC operated manufacturing plant to standard and safe DCs, fixed cost for opening DCs for warehousing, fixed cost for making required arrangement assigning DCs to sales outlet and fixed cost for establishing contractual arrangement with sales outlets to send product from assigned DCs; costs for distributing product from DCs to sales outlets; cost for sending product from plant to retailer in the markets as a part of emergency planning including fixed cost for such arrangement.

$$
\begin{aligned}
TRIDIST \\
= \sum_{p \in P} \sum_{j \in J} \left(\sum_{k \in K} CTR_{pjk'} x_{pjk} + \sum_{k \in K'} xs_{pjk'} \right) \\
+ \sum_{k \in K} (dw_k + sdw) FW_k \\
+ \sum_{m \in M} \sum_{r \in R} \left(\sum_{k \in K} drw_{kmr} + \sum_{k' \in K'} sdrw_{k'mr} \right) FWR_{kmr} \\
+ \sum_{p \in P} \sum_{r \in R} \sum_{m \in M} \sum_{k \in K} CDR_{pkmr} y_{pkmr} \\
+ \sum_{p \in P} \sum_{j \in J} (empp_{pj} FM_{pj} + \sum_{m \in M} \sum_{r \in R} EMD_{pjmr} emgr_{pjmr}) \quad (2.52)
\end{aligned}
$$

Constraint (2.53) imposes integrality.

$$
\begin{aligned}
u_{is}; \, bz_{is} \in \{0,1\}, \, \forall \, i, s; \quad con_{sb} \in \{0,1\}, \, \forall \, b, s; \, em_{ie} \in \{0,1\}, \, \forall \, i, e; \\
dw_k \in \{0,1\}, \quad \forall \, k; drw_{kmr}, sdrw_{k'mr} \in \{0,1\}, \quad \forall \, k, k', m, r, \\
vpp_{pj}, empp_{pj} \in \{0,1\}, \quad \forall \, p, j \quad (2.53)
\end{aligned}
$$

2.4 Numerical Example

In this section, an example consumer product SC case is illustrated to show applicability of the model. The example SC manufactures five products in their five plants and transports them from plants to seven DCs and distributes the products from DCs to five retail outlets in each of their six markets. Each product is made up of five to six components out of total 12 components maintained and procured from a pool of nine quality affiliated suppliers. Two DCs out of the seven are located in safe locations for which there is no past instances of calamity-based disruptions. This section presents analysis of model decisions and outcomes for objective function values. Limited input information is presented when that is needed to explain the model outcomes.

The model plans supply system, manufacturing system, and distribution system resilience and determines coefficient of performance for each system separately and in a combined way to facilitate SC management to understand their resilience status for facing foreseeable and unforeseeable disruptions from natural calamities and manmade problems. This section first discusses various capabilities that model plans for improving *SSCP*, *MSCP*, and *DSCP*. Model outcomes/results for overall coefficient of performances combining SSCP, MSCP, and DSCP and each one separately are presented next along with total cost of the SC.

Before proceeding further, we would like to mention that the model outcomes/results present typical assumed effects of disruptions/calamities by considering random 0/1 inputs for their occurrence and non-occurrences of disruptions, and % effects at scenarios for typical natural and manmade disruptions. The model applied containment measures through the planned absorptive, adaptive, and recovery steps, as applicable. Since absorptive capability can inhibit/contain several foreseeable and unforeseeable disruptions through flexibility, redundancy, partnering, and system-based resilience improvement options, like quality systems, we present them first before going for analysis based on coefficient of performances and total cost.

2.4.1 Model Results for Supply Systems Resilience Coefficient of Performance (*SSCP*)

SSCP considers absorptive, adaptive, and recovery capability for the supply system to comply with customer requirements. As a part of absorptive capability for supply systems the model first evaluated the enlisted suppliers and identified quality affiliated suppliers to prevent quality failure from input supply. Such quality failure may result into a disruption through media attention and product recalls. The model considered eight quality metrics and score for each in a 0–1 scale for each supplier as presented in Table 2.2. The SC for the consumer product set minimum overall quality level to be 0.241 as the qualifying score based on the product of the eight metrics for considering interaction and interdependencies of the metrics. To ensure customer satisfaction, the qualifying score was set considering average minimum score for each attribute to be 0.837 in a 0–1 scale, that resulted $0.837^8 \approx 0.241$. The model could affiliate each of the nine enlisted suppliers based on the overall product value of the eight metrics shown in the

Table 2.2   Selection of quality affiliated suppliers for ordering inputs

| Suppliers | Score by the Suppliers in Each of the **Metrics** | | | | | | | | Overall |
	1	2	3	4	5	6	7	8	Product of metrics
1	0.79	0.95	0.81	0.93	0.82	0.93	0.79	0.9	0.3065
2	0.94	0.95	0.79	0.92	0.82	0.89	0.82	0.93	0.3612
3	0.83	0.89	0.89	0.84	0.91	0.8	0.78	0.79	0.2477
4	0.84	0.81	0.81	0.84	0.87	0.86	0.91	0.85	0.2679
5	0.91	0.91	0.85	0.83	0.82	0.8	0.93	0.81	0.2887
6	0.9	0.92	0.94	0.79	0.94	0.85	0.89	0.8	0.3498
7	0.83	0.89	0.78	0.92	0.91	0.87	0.85	0.81	0.2889
8	0.79	0.93	0.86	0.86	0.9	0.91	0.87	0.89	0.3446
9	0.92	0.83	0.93	0.91	0.8	0.89	0.91	0.85	0.3559

Table 2.3   Typical assignment of **inputs** to suppliers from locations

	1	2	3	4	5
Supplier:	1,2,6	3,7,8	1,3,7,9	2,6	1,2,5,6,9
Suppliers (Locations)	1(1), 2(2), 6(4)	3(2), 7(5), 8(2)	1(1), 3(2), 7(5)	2(2), 6(4)	1(1), 2(2), 5(4), 6(4), 9(5)

last column of Table 2.2. As may be observed in Table 2.2 overall product value resulted 0.2477–0.36. The quality metrics considered for supplier's affiliations are related to supplier's production plant and organizational performances that include: production scrap rate; process capability of the plant; plant reliability based on lean metric OEE; number of training organized and provided to plant operators; employee turnover; number of customer complaints in the last operation year; inventory turnover of the supplier; ISO or any quality management system follower including certifications, for example.

The suppliers for this research are from five locations. The model plans supplier flexibility and supplier's location flexibility by assigning each input to more than one suppliers and suppliers for each input are from more than one location. Table 2.3 presents model decision for such flexibility options in assignment of suppliers.

Based on Table 2.3, the model assigned input 1 to suppliers 1, 2, and 6. According supplier's data supplier 1 is from location 1, supplier 2 is from location 2, and supplier 6 is from location 4. As such input 1 is assigned to three suppliers from three locations, for example.

Other supply system management resiliency options for improving adaptability and recovery capability by initiating emergency measures, business continuity plan, keeping safety stock built in the model by constraints (2.11)–(2.15) have been described in detail in the supply system resiliency planning model. Supply system resilience performance index *SSCP* value is the combined effects of adaptability, recovery, and absorptive capability-based containment options.

2.4.2 Model Results for Manufacturing Systems Resilience Coefficient of Performance (*MSCP*)

Manufacturing system is the most crucial functions of consumer product systems or any manufacturing-based business systems and SCs. Manufacturing system resilience planning includes building absorptive, adaptive, and recovery capability in all the possible ways and options. The model first considers critical to business and critical to quality metrics-based criteria for identifying quality capable plants such that finished goods transported from the plants are of ensured quality. By such steps, SC builds absorptive capability to inhibit quality failure-related disruptions that may create warranty claims, media attention, and product recalls. It then plans manufacturing capacity flexibility by assigning production of product to more than one plant, creating pooling options from the overall plant capacity and sister company plants where applicable; creating decentralized capacity options through external vendors which will provide absorptive capability through flexible capacity, and adaptive capability to provide support when central plant fails. It also builds network-based partnering relationships to obtain product support through mutual benefit-based contractual arrangement to contain disruptions of different severity levels through absorptive and adaptive strategy.

The model considers nine critical to business and critical to quality attribute metrics for plant capability determinations. For each of the attributes, the plant is evaluated in a 0–1 scale. A plant is considered quality capable based on its obtaining of higher than minimum qualifying score for the combined effects of all nine metrics (taking their product). The minimum qualifying score for the combined effect is set by the SC. For this example, consumer product SC, the attributes considered include: scrap rate/rejection rate; process capability; overall equipment effectiveness; number of loss time accidents for the last year; number of customer complaints linked to the product from the plant; number of lean and sustainability-based

Table 2.4   Plant capability determination

Plant	Score by the Plants for Each of **Critical to Quality and Business Metrics**									Overall	Capable Y/N
	1	2	3	4	5	6	7	8	9		
1	0.77	0.66	0.87	0.78	0.85	0.89	0.78	0.62	0.75	0.0946248	Y
2	0.81	0.87	0.74	0.76	0.62	0.88	0.8	0.69	0.72	0.08594	Y
3	0.89	0.69	0.69	0.86	0.6	0.64	0.85	0.72	0.67	0.0573778	Y
4	0.79	0.64	0.79	0.88	0.62	0.82	0.68	0.76	0.89	0.082193	Y
5	0.75	0.6	0.82	0.65	0.83	0.63	0.63	0.63	0.74	0.0368359	N

Table 2.5   Allocation of production to plants

	Allocation of **Products** to Plants				
	1	2	3	4	5
Assigned plants	2,3,4	2,3,4	1,2,3,4	1,2,3	2,3,4

training provided to operator; ISO9000 and ISO14000 following company; lean six sigma following company. The minimum qualifying score for the combined effect was set to be 0.057 based on the considerations that each attribute should be at an average: 0.7275 ($0.7275^9 \approx 0.057$). The model found four plants (1, 2, 3, and 4) to be quality capable but not plant 5. Table 2.4 presents the plant capability determination process by the model. Based on score for each metric by plants: 1, 2, 3, 4, and 5, overall combined score of the nine metrics by the plants {0.0946; 0.0859; 0.0574; 0.0822; 0.0368} as shown in the last column. Since overall combined score for plant 5: 0.0368 <0.057, the model could not decide plant 5 to be capable, for example, as may be observed in the last column of Table 2.4.

Table 2.5 presents model decision for allocating production of products to plants. As may be observed, the model allocated three to four capable plants for producing each product. For an example product 1 is allocated to plants 2, 3, and 4. As discussed before such allocation provides capacity flexibility and builds absorptive capability for containing disruption of a plant by taking support from nonaffected flexible capacities.

Other resiliency options through adaptive and recovery steps, such as product support from external vendors, network partners; using flexible capacity through pooling sister plants and SC operated plants as included in the model by constraints (2.23)–(2.27) will work with absorptive capability through flexibility and quality assurance resilience to keep the overall coefficient of performance (*MSCP*) at the desired level.

2.4.3 Model Results for Distribution Systems Resilience Coefficient of Performance (*DSCP*)

Similar to supply system and manufacturing system resilience planning distribution system resilience planning includes: absorptive, adaptive, and recovery strategy for containing disasters and disruptions faced by product distribution system. It creates absorptive capability by allocating more than one standard DCs and one safe DC for distributing product to a retailer. Table 2.6 presents DCs allocated to retailer for creating distribution flexibility. Such flexibility ensures product distribution to a retailer if one of the DC allocated to the retailer is disrupted due calamity, or cannot distribute product due to border congestion, custom strike, big accident, or similar disruption events. According to Table 2.6 the model allocated each of the five DCs to each retailer in each market. This happened complying with flexibility creation constraint (2.35) and capacity limitation of DCs by constraint (2.31). Table 2.6 also presents typical allocation of safe DCs (mentioned as SDC in the table) to retailer in each market. For an example, safe DC 6 (shown as SDC 6) is allocated to retailers 3 and 4 in market 1 but other retailers are allocated safe DC 7.

The model includes adaptation and recovery resilience strategies for distributing product to retailers are: pooling product from partially disrupted and non-disrupted DCs; taking emergency product supply to markets from external vendors; and emergency support through direct supply of product from plant to retailers when DCs are disrupted. Such strategies are included in the

Table 2.6 Allocation of distribution centers (DCs) to retailers in markets by the model

	Allocation of Distribution Centers (DCs) to **Retailers**				
Market	1	2	3	4	5
1–6	DC1–DC5	DC1–DC5	DC1–DC5	DC1–DC5	DC1–DC5
	Typical Allocation of Safe DCs (SDC) to **Retailers** in markets				
1	SDC7	SDC7	SDC6	SDC6	SDC7
2	SDC7	SDC7	SDC7	SDC6	SDC7
3	SDC7	SDC7	SDC6	SDC6	SDC7

model through constraints (2.39)–(2.44), which work in a combined way with absorptive strategy to keep overall distribution coefficient of performance *DSCP* at a desired level.

2.4.4 Analysis of Resilience Coefficient of Performances and Overall SC Cost

Since resilience coefficient of performances for the functional systems and overall SC are needed to be planned within the optimum cost for the desired overall business performance, we solved the model to obtain pareto optima solutions for resilience performance metrics vs total cost of SC for the example case. Such solution is planned for providing managerial insights to SC management such that SC may decide *OVCP* level at which they would like to operate, given the requirements of investments in the overall SC cost. Table 2.7 presents pareto optima solutions for the overall resilience performance index *OVCP* vs total costs of SC (*TC*). Out of seven solutions presented in Table 2.7, solutions 1 and 7 are not pareto optima solutions. They are independent for two objective functions.

Since objective Function 1 *OVCP* has the highest value of 2.772 in solution 1 and lowest value 2.715 for minimizing SC cost in solution 7, we divided difference of (2.772 – 2.715) by 5 to obtain five pareto optima solutions for model definitions in the last column of Table 2.7. As may be observed in Table 2.7 as the *OVCP* value increased from 2.715 for solution 7 to 2.772 for solutions 1 and 2 total SC cost increased from $95.857 Million to $96.787 Million. This increased cost $ (96.787 – 95.857) = $0.93 million is spent in creating resilience by the model. The increment in cost is only 0.97% but improvement in resilience is (2.772 – 2.715) * 100/2.715 = 2.09%. By this improvement in resilience coefficient performance SC will be able to contain several disruptions considered in the study. The increment in *TC* and its influence on investment from such incremental cost may also observed in Figures 2.1 and 2.2. In addition to the data for total SC costs (*TC*) and overall resilience performance coefficient performance (*OVCP*) Table 2.7 also presents resilience coefficient of performance metric values for supply systems (*SSCP*), manufacturing systems (*MSCP*), and distributions systems (*DSCP*) functions. These values provide further managerial insights. For example, with the changes in investments and overall resilience increases, but there is minimal to nil change in the supply system resilience performance index and supply system total operation cost.

Table 2.7 Pareto optima solutions for the overall resilience performance metrics vs total costs

Soln.	OVER RESCP (OVCP)	SSCP	MSCP	DSCP	Total cost (TC) $ Million	Supply System Cost $ Million	Manuf. System Cost $ Million	Distri. Systems Cost $ Million	Model Definition
1	2.772	0.908	0.980	0.884	103.80	53.09	45.158	5.554	*Max OVCP*
2	2.772	0.908	0.980	0.884	96.787	51.036	40.615	5.136	*Min. TC, s.t. OVCP = 0.2.772*
3	2.761	0.909	0.975	0.878	96.002	51.036	39.637	5.329	*Min. TC s.t. OVCP = 0.2.761*
4	2.749	0.909	0.974	0.866	95.942	51.036	39.637	5.269	*Min. TC s.t. OVCP = 2.749*
5	2.738	0.909	0.974	0.855	95.902	51.036	39.637	5.229	*Min. TC s.t. OVCP = 0.2.738*
6	2.725	0.909	0.974	0.842	95.870	51.036	39.637	5.197	*Min. TC s.t. OVCP = 0.2.725*
7	2.715	0.908	0.975	0.832	95.857	51.036	39.637	5.184	*Min. TC*

It may be mentioned here even at the basis level performance, based on which we are comparing increment or decrement for SC costs and *OVCP*, SC needs necessary inputs and supply management system with desired resilience performance for fulfilling market requirements for realizing required product in the manufacturing systems. It may be also noted that no demand increase is considered in the planned period under considerations. It is rather reverse, requirements of inputs, may be reduced due to disruptions and failures in the manufacturing and distribution systems. So, for the supply system function maintaining *SSCP* at the desired level is the most challenging one. By including various supply system resilience measures this study planned to keep the *SSCP* at the desired performance level which is depicted by the almost steady value of *SSCP* for pareto optima solutions as shown in Table 2.7.

Table 2.7 also includes resilience coefficient of performance index metrics for manufacturing systems (*MSCP*) and that of distribution systems (*DSCP*). Since *SSCP* is almost steady for the pareto optima solutions, changes occur in *MSCP* and *DSCP* for the increments and decrements in *OVCP* in the pareto optima solutions. For discussing and analyzing changes in the *MSCP* and *DSCP*, we included them in the *OVCP* vs *total cost* charts as shown Figures 2.1 and 2.2.

As is evident in Table 2.7 and confirmed by Figures 2.1 and 2.2; both *MSCP* and *DSCP* increase with each increment in *OVCP* through increase in the investment as depicted by total SC cost values. We may also observe that increase in *DSCP* is gradual with the increments in *OVCP* values but *MSCP* increases in three steps and reaches peak. It first increases from 0.974 to 0.975 and in the next jump reaches its peak 0.980. In the case of *DSCP*, its increments are gradual from 0.832 to 0.878 and then it reaches its peak 0.878–0.884. Such trend is obvious, improvements in manufacturing system resilience comes mostly through resource-based supports/investments, such as pooling capacities of company operated and sister company plants, or though creating contract with external vendor, and finally from network partner plants through formal contract. Since pooling of capacity seldom happens, such resilience also needs investments to plan and create system to accumulate product and transport to DCs to fulfill market requirements. Whereas distribution system resilience steps are creation of systems or provisions with the resources bases or sources already created by the SC (such as obtaining product by pooling partially affected or nonaffected DCs) when each retailer is allocated for almost all DCs, minimum investment effort is needed to obtain such supports. It may also be noted that such support may

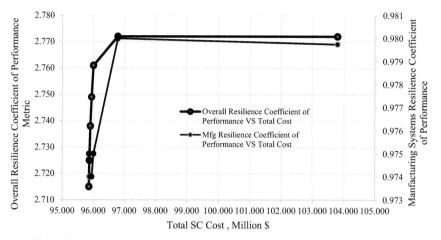

Figure 2.1 Overall resilience coefficient performance vs total cost with *MSCP*.

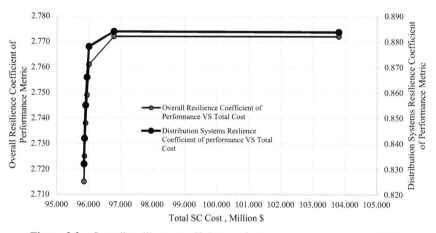

Figure 2.2 Overall resilience coefficient performance vs total cost with *DSCP*.

not necessarily improve performance indices significantly, given the situation that it takes support only up the extent it needs without much investment. Similar is the case of getting direct supply of products from plants when other resilience support fails.

As evident from the above analysis that the model effectively planned supply system resilience performance (*SSCP*), manufacturing system resilience performance (*MSCP*) and distribution system resilience performance (*DSCP*) at an optimal level for fulfilling market requirements within optimum cost.

The pareto optima solutions planned for decision process provides options to managers to select suitable coefficient performance index level with planned cost in a what-if analysis situations. One significant finding in this research is that after a certain *OVCP* level (such as 2.762 for solution 2) investment cannot influence *OVCP* improvement (please see the chart). This finding provides managerial insights that managers should follow resilience improvement strategy to achieve their desired level and after that they should concentrate on monitoring news media, social media, national weather broad casting for emerging disasters and new types of threats.

2.5 Conclusions

This research contributed by creating a new resilience system design model for consumer product system. The model is built taking a system approach such that it can be applied in similar systems or expanded to any other systems and SCs. It planned supply, manufacturing, and distribution systems management functions for a SC integrating absorptive, adaptive, and recovery capability in the planning process to make the system resilient. Such integration provides managerial guidelines and insights on the critical points on which SC needs to pay their attention from the beginning of their planning to make their system resilient. It will systematize the overall SC operation and control for each containment measures in supply, manufacturing, and distribution systems to make the SC always prepared for facing disruptions and disasters.

The research formulated resilience performance coefficient for each SC functions and connected them with business systems planning such that resilience performance is targeted to fulfill SC business performance requirements by complying with market requirements.

The research also provides detailed ideas on the disaster occurrence-related inputs needed through monitoring and following-up information network, national, and international news media on weather updates and emerging natural calamities, and other threats.

Since resilience creation needs investment, the research includes ideas on cost implications for each resilience steps. The research also includes pareto optima solutions for overall resilience coefficient performance of the SC vs total SC cost such that SC management may plan their desired resilience performance level by taking a what—if approach. It will facilitate gradual planning of investment and resilience performance improvement options based on their targets.

The study created future research scope for expanding such resilience system creation approach and its practical application in an automobile SC or other durable consumer goods.

References

Bhamra, R., Dani, S. and Burnard, K. (2011). Resilience: The concept, a literature review and future directions. *International Journal of Production Research*, 49(18), 5375–5393.

Bradley, J. R. (2014). An improved methodology for managing catastrophic supply chain disruptions. *Business Horizons*, 57(4), 483–495.

Chopra, S. and Sodhi, M. S. (2014). Reducing the risk of supply chain disruptions. *MIT Sloan Management Review*, 55(3), 72–80.

Chopra, S. and Sodhi, M. S. (2004). Managing risk to avoid supply chain breakdowns. *MIT Sloan Management Review*, 46(1), 53–62.

Das, K. (2011). A quality integrated strategic global supply chain model. *International Journal of Production Research*, 49(1), 5–31.

Fiksel, J., Polyviou, M., Croxtom, K. L. and Pettit, T. J. (2015). From risk to resilience: Learning to deal with disruption. *MIT Sloan Management Review*, 56(2), 78–86.

Hanson, J. D., Melnyk, S. A. and Calantone, R. A. (2011). Defining and measuring alignment in performance management. *International Journal of Operations & Production Management*, 31(10), 1089–1114.

Holling, C. S. (1973). Resilience and stability of ecological systems. *Annual Review of Ecological Systems*, 4(1), 1–23.

Hussaini, S., Khaled, A. A. and Sarder, M. D. (2016). A general framework for assessing system resilience using Bayesian networks: A case study of sulfuric acid manufacturer. *Journal of manufacturing Systems*, 41, 211–227.

Leveson, N.(2004). A new accident model for engineering safer systems. *Safety Science*, 42(4), 237–270.

Li, Y., Zobel, C. W. and Russell, R. S. (2017). Value of supply disruption and information inaccuracy. *Journal of Purchasing and Supply Management*, 23(3), 191–201.

Linkov, I., Trump, B. D. and Fox-Lent, C. (2017). Stemming flirtations with disaster, resilience could beat risk management in dealing with ecological threats and natural catastrophes. *ISE Magazine*, 49(3), 32–37.

Mackenzie, C. A and Hu, C. (2018). Decision making under uncertainty for design of resilient engineered systems. *Reliability Engineering and Systems safety*, in press, https://doi.org/10.1016/j.ress.2018.05.020.

Madani, A. M. and Jackson, S. (2009). Towards a conceptual framework for resilience engineering, *IEEE Systems Journal*, 3(2), 181–191.

Munoz, A. and Dunbar, M. (2015). On the quantification of operational supply chain resilience. *International Journal of Production Research*, 53(22), 6736–6751.

Nan, C., and Sansavini, G. (2017). A quantitative method for assessing resilience of interdependent infrastructures. *Reliability Engineering and System Safety*, 157, 35–53.

NIAC. (2009). Critical Infrastructure Resilience Final Report and Recommendations, US Department of Homeland Security, National Infrastructure Advisory Council, Washington, DC.

NIPP. (2009). 2009 National Infrastructure Protection Plan, US Department of Homeland Security, Washington, DC, available at http://www.dhs.gov/files/programs/editoral_0827.shtm

National Oceanic and Atmospheric Administration (NOAA). (2017). About NOAA. http://www.noaa.gov/about-noaa.html, accessed in June 2017.

Park, J., Seager, T. P., Rao, P. S. C., Convertinao, M. and Linkov, I. (2013). Integrating risk and resilience approaches to catastrophic management in engineering systems. *Risk Analysis*, 33(3), 356–367.

Raj, R., Wang, J. W., Nayak, A., Tiwari, M. K., Han, B., Liu, C. L. and Zhang, W. J. (2015). Measuring the resilience of supply chain systems, using a survival model. *IEEE Systems Journal*, 9(2), 377–381.

Sheffi, Y. and Rice Jr, J. B. (2005). A supply chain view of the resilient enterprise. *MIT Sloan Management Review*, 47(1), 40–48.

Sheffi, Y. (2015). Preparations for disruptions through early detection. *MIT Sloan Management Review*, 57(1), 30–41

Shen, L. and Tang, L. (2015). A resilient assessment framework for critical infrastructure system. The *First International Conference on System Engineering*, (RP0290, 2015 ICRSE), 1-5, DOI: 10.1109/ICRSE.2015.7366435.

Simchi-Levi, D., Schmidt, W. and Wei, Y. (2014). From superstorms to factory fires, managing unpredictable supply chain disruptions. *Harvard Business Review*, January-February, pp. 96–101.

Soni, U., Jain, V. and Kumar, S. (2014). Measuring supply chain resilience using deterministic model approach. *Computers & Industrial Engineering*, 74, 11–25.

The National Academy Science (2012). Disaster resilience: A national imperative. Publisher: National Academies Press.

United Nations International Strategy for Disaster Reduction. (2013). Synthesis report: Consultations on a post-2015 framework on disaster risk reduction (HFA2). UNISDR, Geneva.

Webb C. T. (2007). What is the role of ecology in understanding ecosystem resilience? *Bioscience*, 57(6), 470–471.

Yilmaz-Borekci, D., Say, A. I. and Rofcanin, Y. (2015). Measuring supplier resilience in supply networks. *Journal of Change Management*, 15(1), 64–82.

Youn B. D., Hu, C. and Wang, P. (2011). Resilience-driven system design of complex engineered systems. *Journal of Mechanical Design*, 133(10), 101011(15).

Zhang, W. J. and Lin Y. (2010). On the principle of design of resilient systems: Application to enterprise systems. *Enterprise Information Systems*, 4(2), 99–110.

3

Definitions of Resilience and Approaches for Mathematical Modelling of Its Various Aspects

Jaqueline Hemmers[1,2], Matthias Winter[3], Stefan Pickl[1,2,*], Markus Gerschberger[3] and Benni Thiebes[1,2]

[1]German Committee for Disaster Reduction DKKV, Bonn 53113, Germany
[2]COMTESSA, Universität der Bundeswehr München, Neubiberg 85577, Germany
[3]LOGISTIKUM, University of Applied Sciences Upper Austria, Steyr 4400, Austria
E-mail: jaqueline.hemmers@dkkv.org; Matthias.Winter@fh-steyr.at; stefan.pickl@unibw.de; Markus.Gerschberger@fh-steyr.at; benni.thiebes@dkkv.org
*Corresponding Author

This paper presents the etymology of the term resilience and examines the different approaches to it. In this context, a closer look is taken on the definitions and concepts of engineering resilience and ecological resilience.

3.1 Introduction

In 2017, more than 95.6 million people were affected by natural disasters (World Security Report, 2018). Natural as well as human-made disasters, such as the 2011 Tohoku earthquake and tsunami, or Hurricane Florence in 2018 often disrupt societal processes and public life, shaking our image as a "disaster resilient society."

Given the complexity and interdependence of modern risks, it is impossible to hedge against all potential events (Bara, 2010). Whether natural disasters and extreme weather events result in a disaster depends on the vulnerability, exposure, and the ability of the population to cope with disturbances and recover from their effects (Aitsi-Selmi et al., 2015).

Consequently, concepts of resilience, vulnerability, adaptation, and sustainable development have become center points of attention in research and policy (Christmann et al., 2011; Cutter, 2016, O'Hare and White, 2013). They led to a variety of definitions and mathematical concepts.

While vulnerability points toward the need for a system to change in order to become less affected by disasters, the concept of resilience has a more positive focus (Sudmeier-Rieux, 2014), resilience puts a stronger emphasis on the abilities of individuals, communities, societies, and even infrastructure to deal with disasters. Therefore, resilience empowers self-reliance and allows a more holistic, integrated, and collective approach (Manyena et al., 2011), which can be applied to various disciplines and contexts (Alexander, 2013).

Today, the term resilience is used in various disciplines, leading to multiple and often contradictory definitions (Sudmeier-Rieux, 2014). In 2013, resilience was even declared as the buzzword of the year by the Time magazine (Pugh, 2014).

Besides giving a general overview on the etymology of the term resilience, this paper presents different approaches to resilience. These include definitions of engineering resilience, ecological resilience, and socioecological resilience.

3.2 State of Research

The word resilience has a long history (Alexander, 2013). It originates from the Latin word "resilio" or "resiliere" which translates to "jumping back" or "bouncing back" (Kharrazi 2018, Sudmeier-Rieux, 2014). The beginning of resilience research is attributed to different authors, depending on the discipline and the national localization (Gabriel, 2005). According to Yunes (2003), Amorim (2009), or Sudmeier-Rieux (2014) the word resilience was first used around 1807. Back then, the English physicist Thomas Young described a material's capacity to absorb energy without suffering permanent deformation (Sudmeier-Rieux, 2014).

Since then, the usage of the term resilience has grown greatly (see Figure 3.1). In the 1950s, the term was used not only by mechanics, manufacturing, and medicine, but also in ecology and social sciences (Alexander, 2013).

Other intellectual roots can be traced back to psychology and child behaviour (van der Vegt et al., 2015). Within psychology, the term resilience was used to describe the individual's or the collective's capacity to adapt despite threatening circumstances or shocks (Detten et al., 2013; Sudmeier-Rieux, 2014 and Lewis, 2013).

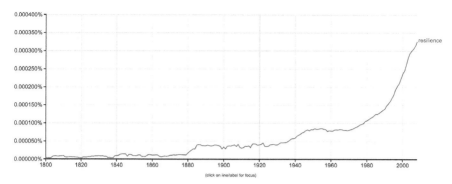

Figure 3.1 Usage of the term resilience from 1800 to 2018 in books over the selected years (Ngram Viewer, https://books.google.com/ngrams).

The concept of resilience also became popular within the field of engineering as well, focusing on human constructions (Weichselgartner and Kelman, 2015).

Through various essays on resilience and ecological systems, Holling strongly influenced the conceptual understanding of resilience (Goecke, 2017). In his first essay, "Resilience and stability of ecological systems," Holling stated that systems at risk of crisis are forced to adapt to changes. Two notions of resilience were introduced: The first one, called *engineering resilience*, refers to *"the ability of an ecosystem to return to stability or equilibrium after a disturbance."* The second one, *ecological resilience*, refers to the ability of systems *"to absorb changes (...) and still persist"* (Holling, 1973).

For a comparison between both notions of resilience, see Figure 3.2.

Today resilience is being used in many disciplines and fields. These include, e.g., share marketing, material science, psychotherapy, politics, sociology, social science, biology, geography, design, and community development (Lewis and Kelman, 2010; Alexander, 2013; White and O'Hare, 2014; Pugh, 2014; Adger, 2000; Holling, 2001).[1] The application in different disciplines led to a variety of definitions, meanings and focus (Weichselgartner and Kelman 2015, Alexander 2013, Christman and Ibert, 2015). Despite their different backgrounds, notions of resilience share common threads, such as the *"ability of materials, individuals, organizations and entire social-ecological systems, from critical infrastructures to rural communities, to*

---

[1]For a detailed comparison of various definitions of resilience, see Manyena (2006), Sudmeier-Rieux (2014).

	Engineering resilience	Ecological resilience
Definition of resilience	Elasticity which emphasizes resistance to disruption and rate at which a system returns to a single steady-state or cyclic state following a perturbation	Amount of change or disruption that is required to transform a system from being maintained by one set of mutually reinforcing processes and structures to a different set of processes
Focus	Efficiency, control, constancy predictability	Persistence, adaptiveness, variability, unpredictability
Perspective	Fail-safe design, optimal performance	Evolutionary and development perspective
Equilibrium	Stability near an equilibrium steady-state	Conditions far from any equilibrium steady-state, instabilities can flip a systems into another regime of behaviour
Measure of resilience	Resistance to disturbance, speed of return to the equilibrium	Magnitude of disturbance that can be absorbed before the system changes its structures by changing the variables and processes that control behaviour
Stability focus	Maintaining efficiency of function	Maintaining existence of function
Spatial scale	Simplified and untouched ecological systems, limited scale to small enclosures	Landscapes, population interacting in nature
System objective	System with a single operating objective, goal is sustained production	Complex system with multiple objectives
Basic assumption	Global stability, best equilibrium steady-state	Flips from one operating state to another cannot be avoided

Figure 3.2 Comparison between engineering and ecological resilience (Edited, after Holling, 2008).

withstand severe conditions and to absorb shocks" (Weichselgartner and Kelman, 2015).

Thus, resilience is used to explain the functioning of an existing system during and after disturbances (MacKinnon and Derickson, 2013). Many authors however argue that resilience is much more than "bouncing back" to the baseline. Instead, it would be desirable that communities dealing with natural hazards would not just return to the status quo (Goecke, 2017). Consequently, other authors such as Birkmann (2008) have taken different approaches to define resilience.

The first one, the socioecological resilience, views resilience as a system which has the ability to learn from shocks. The second one views resilience as a part or a counterpart of vulnerability.

Socioecological resilience puts a strong emphasis on the human-environment system, focusing less on the robustness of systems. Instead, the system's ability to self-organize, learn, and adapt to changes is stressed (IPCC, 2012). The definition of the United Nations Office for Disaster Reduction (UNISDR) in 2009 can be stated as one example which describes socioecological resilience:

> *"the ability of a system, community or society exposed to hazards to resist, absorb, accommodate to and recover from the effects of a hazard in a timely and efficient manner, including through the preservation and restoration of its essential basic structures and functions."* (UNISDR 2009 in Alexander, 2013)

The school of social-ecological resilience considers systems as extremely dynamic. Each system is subject to a cycle of growth, conservation, decline, and reorganization.

The term resilience here refers to the characteristics or capacities of a social, ecological, or socioecological system which enable this system to sustain essential functions under shocks or to quickly shut down interruptions in system processes or functions (Birkmann, 2008), represented in the concept of panarchy. Through changes, planning and recovery, systems can mitigate the effects.

Conceptually, the terms resilience and vulnerability are closely related. According to Manyena et al. (2011) both concepts are often treated similarly. Different authors therefore display resilience as a part or counterpart of vulnerability (Goecke, 2017, Manyena, 2006) as *"both are different manifestations of a variety of response processes to change"* (Weichselgartner and Kelman, 2015).

In the former, resilience is associated with coping, reacting, and adapting to changes. Sudmeier-Rieux (2014) instead argue that resilience is the opposite, positive side of vulnerability. Overlaps between the two concepts exist, such as demographic, social, cultural, economic, political, etc. factors.

Poor households, however, may be vulnerable but at the same time resilient as well (Sudmeier-Rieux, 2014, Kelman et al., 2016).

3.3 Mathematical Aspects of Resilience

Consider now a system in a steady state of performance. Hardly any system is independent of outside influences, which may disrupt the system with different intensities and have different effects on it. However, those influences may change the steady state of the system and the managers of the system may try to counter this change if it has a negative effect on performance. It is reasonable to ask about the ability of the system to resist such changes and to recover after such a disruption, which leads to the development of the term resilience, discussed above.

In this paragraph, we consider systems whose states can be described via a number of properties. Each property can be quantified by a single value, leading to a description of the state of the system being an n-dimensional vector at each point in time. We are interested in the change of the system's state – referred to as the behaviour of the system over time. Such a system can be modelled mathematically using the theory of dynamical systems, where a function is used to describe the change of the system's state over time. Starting at any point in the n-dimensional space, the function describes, where the state of the system will end up after a positive time. Formal, a dynamical system is given by:

$$x'(t) = f(x, t), \tag{3.1}$$

where $x'(t)$ indicates the change of the system's state, starting at point x, after time t. The first definitions of resilience in this text are based on the assumption that the system can find itself in a steady state. Whether this state is stable or not is a different question, but the assumption that a system can be in a steady state for a certain period of time is also questionable, and hence a constraint. We will come back to this thought later.

A steady state is referred to as an equilibrium and the question of stability of an equilibrium can arise. Stability theory gives different notions of stability (Jen, 2003) – i.e., the stability of the state of a dynamical system and the stability of the structure of a dynamical system. A state is said to be stable if small perturbations of one or more properties lead to a new system state, which stays close to the original solution for all time. For example, consider a railway system, where the delays of all trains are zero.

If one train is slightly delayed due to some outside influence, this delay should only lead to small delays of other trains and not to an escalation of delay time for any of the trains. For structural stability, a small change of the system itself has to be considered, which can be expressed as a small change of the function f, which was introduced above.

The system is structurally stable if any small change of this function leads to a new dynamical system that has qualitatively the same dynamics as the original system. Using the same example as before, the small change of the system could be a slightly different velocity of the trains, which should not result in the disappearance of an equilibrium or emergence of another one.

To get a clear understanding of the model, we point at the two different origins of changes of the system's states. The function f covers only changes that are based on the system's behaviour, meaning that the system's natural manner causes the change. Talking about resilience, it is of additional interest what happens if there is a change that lies not in the natural behaviour of the system, for example, a disruption or a change of the original conditions of the system. Such a change may just move the state of the system to a different point, where the system starts out again and moves back to an equilibrium or not.

Otherwise, such a change can move the system to a different point and in addition changes the natural behaviour of the system, i.e., the function f, which describes the behaviour of the system. Next, we consider the different terms of resilience that were already discussed above.

3.3.1 Engineering Resilience

Holling and Crawford (1996) compare two concepts of resilience: The more traditional *engineering resilience* is measured by the resistance to disturbance and the speed of return to the equilibrium. These two attributes are important characteristics of resilience, though not the only ones, as we will see later on. However, engineering resilience describes the process of a system, reacting to deviations near an equilibrium or steady state and being able to return to that equilibrium. Walker et al. (2004) define resistance – only one aspect of resilience – as "the ease or difficulty of changing the system." The addendum "to disturbance" is crucial for resistance as an aspect of resilience, since one would require a low resistance to improvements and only a high resistance to disturbance. Intending more clarity, we call those two aspects *resistance to disturbance from equilibrium* and *speed of return to equilibrium*.

3.3.2 Ecological Resilience

If the system operates at a long distance from the steady state, deviations are likely to have larger consequences, turning the system to a different *regime of behaviour*. This different regime of behaviour would be modelled as a different function f, which describes the behaviour of the system.

Here, resilience is defined as "the magnitude of disturbance that can be absorbed before the system changes its structure" (Holling and Crawford, 1996) – referred to as *ecological resilience*. While engineering resilience considers only one equilibrium, ecological resilience allows for more. However, it is not said that the new regime of behaviour results in bad performance. Wang et al. (2017) translated this definition into a quantitative one, using the original–stable–state of the system x_0 and denoting the state after a disturbance by x_{Md}, the magnitude of disturbance is given by

$$Md = x_0 - x_{Md}. \tag{3.2}$$

The maximal magnitude of disturbance is reached by x_{Md_min}, which is the worst state the system can be before it changes its structure and turns to a different regime of behaviour. It is also notable that Md is monotonically decreasing in x_{Md} (Wang et al., 2017).

According to the paper of Walker et al. (2004) resilience has four important aspects.

Defining resilience as "the capacity of a system to absorb disturbance and reorganize while undergoing change so as to still retain essentially the same function, structure, identity, and feedbacks," latitude, resistance, precariousness, and panarchy are introduced. Latitude of resilience refers to the maximal intensity of disruption the system can take without losing its ability to recover.

This strongly reminds us of the concept of ecological resilience, where the magnitude of disturbance before changing the structure of the system was considered. The ease or difficulty of changing the system is named resistance of the system. We discussed this aspect already with engineering resilience. It influences the extent of change from the steady state of the system being exposed to a disruption. The larger the resistance, the smaller the change will be. Additionally, with a large latitude the system will be able to recover from larger changes, compared to a system with smaller latitude. The third aspect, precariousness, refers to the distance to a point, where recovery is not possible any more. At the equilibrium, the precariousness is equal to the latitude, and decreases if the system deviates from this steady state.

Revisiting the quantitative definition of Wang et al. (2017), precariousness is the additive inverse to the magnitude of disturbance and both sum up to the latitude, being the maximal magnitude of disturbance. The fourth aspect, panarchy, is not considered here, since it only depends on the state of the subsystems or supersystems at different scales to the scale of the system of interest.

So far, the definitions of resilience assume that the goal is the return to normality, assuming that such an equilibrium exists. In ecological resilience, the existence of more than one equilibrium is assumed, but which equilibrium is the best? This thought leads to evolutionary resilience, well described by (Davoudi et al., 2012) and lies beyond the scope of this text.

Summarizing, we have found the following aspects of resilience:

- resistance to disturbance from equilibrium
- speed of return to equilibrium
- the magnitude of disturbance that can be absorbed before the system changes its structure
- latitude of resilience
- precariousness
- panarchy

Vugrin et al. (2011) considered additionally the amount of extra resources that are needed to recover after a disruption. They defined resilience as "the ability to reduce efficiently both the magnitude and duration of deviation from targeted system performance." A quantification of this definition was again given by Wang et al. (2017), extending their formula for ecological resilience:

$$R_{System} = \frac{x_0 - x_{Md}}{T} + \frac{1}{M_{Resource}}, \qquad (3.3)$$

where T is the time used for recovery and $M_{Resource}$ is the amount of extra resources used for recovery. Again the monotonicity of this term of resilience is given; decreasing with T, $M_{Resource}$, and x_{Md}. Less time means quicker recovery, less extra resources mean more efficient recovery and lower x_{Md} means a higher magnitude of disturbance, caused by the disruption.

The first three aspects of Walker et al. (2004) – latitude, resistance, and precariousness – can be expressed using the terms basins of attraction and stability landscapes. Hence those terms are introduced shortly.

An *attractor* is a state represented by a point, where all points in the neighborhood of that point tend to that attractor and end up there as time increases to infinity. Hence, an attractor is an equilibrium.

The converse is not always true, but if the equilibrium is asymptotically stable, then it is also an attractor. All the points that tend to the attractor are summarized as the *basin of attraction*.

Following the ideas of ecological resilience, there can be more than one attractor and a system staying in one basin of attraction could lose its ability to return to that attractor, if exposed to a disruption.

The system may then tend to a different attractor with a different basin of attraction. The function f, describing the behaviour of the system, may have different values in respective basins, accounting for the new regime of behaviour. Considering different attractors and therefore different basins of attraction, the stability landscape is formed by those basins and the boundaries that separate them in the way that crossing a boundary, the system will tent to the new attractor.

3.3.3 Attractor-Based Resilience

Following Martin et al. (2011), we start with the attractor-based definition of resilience. In the theory of dynamical systems, resilience can be defined in two ways. First, as the reciprocal of the return time to the equilibrium. In the example of the railway system, a failure of the power supply may lead to delays of some trains. The inverse of the time that is needed to restore the desired function, i.e., without delays, would measure the resilience in this case. Note, that this measures the resilience of a point, that represents a possible state of the system. Moreover, unless this value of resilience is not zero, counting for infinite time of recovery, the basin of attraction is not changed.

States of zero resilience can have positive resilience with respect to a different attractor; hence this definition depends on the considered attractor.

The second definition uses the size of the basin of attraction. While Walker et al. (2004) talk about latitude of resilience as the size of the basin of attraction, Martin et al. (2011) define resilience as the distance from the equilibrium state to the boundary of the basin of attraction. This measure of resilience corresponds to the attractor of the respective basin where the system is stated. Considering several attractors, the resilience values must be set in relation to values of resilience of other basins of attraction. However, this approach does not include the state of the system, unless the system is in a state of equilibrium.

Both definitions do not allow the consideration of management action as a counterpart to an ongoing disruption. Therefore, a viability-based definition of resilience is given, first without management action and second with.

3.3.4 Viability-Based Resilience

For the details of viability theory, see Aubin (1990), who further contributed to the development of the theory and also investigated interesting applications for viability theory recently (Aubin and Désilles, 2017). The goal of viability theory is to keep a system at the point of the state space where it can survive. This idea can be used to define resilient points in the state space as points that enable a recovery of the system, as an alternative approach compared to the attractor-based definition using return time.

To start, we give definitions of the viability kernel and the capture basin. The viability kernel of a set of states A is defined as the set of points, which stay in that set A forever, i.e., for all positive times t.

$$Viab_f(A) = \{x \in A : f(x,t) \in A \; \forall t \geq 0\}. \qquad (3.4)$$

Complementary to the viability set, the capturing set of a set of states B, captures all states of the state space, that will reach B after a finite time T.

$$Capt_f(B) = \{x \in B : \exists T > 0 : f(x,T) \in B\}. \qquad (3.5)$$

The points of the capturing set will reach a target set B at some finite time, while the points of the viability kernel will stay in a target set A for all time. The set of states, where one wants the system to be depends on the respective application. For example, the set of desired states can be derived from a certain level of performance. For the railway system, it will not be necessary that the scheduled times are met exactly, rather it will be desired that the scheduled times hold plus minus a few seconds.

Using this concept, a resilient point can be defined as a point that reaches the set of desired states in a finite time and stays there forever. This is given by the capturing set of the viability kernel of the set of desired states. We shall call this set the basin of resilience. A state, where the scheduled times cannot be recovered any more, would have resilience zero in this definition.

Comparing this concept to the previous attractor-based definition, the viability kernel of a set of desired states would be the intersection of the basin of attraction and the set of desired states. The inverse of the time that is needed to reach the viability kernel is associated with the resilience of the corresponding state. Hence, points of the viability kernel itself have infinite resilience. Note that this definition does not need the concept of attractors, since the set of desired states is the base of the resilience basin. However, if this set is assumed to be the set of good attractors, both concepts lead to the same resilient states, i.e., the resilience basin equals the attraction basin in this case. In contrast, the values of resilience, i.e., the return time to the desired states, can always be different comparing both definitions.

For applications, the set of desired states has to be determined by the respective goal of the application, which is a practical advantage. As mentioned above for the railway system, the set of desired states will often include a small corridor of the exact desired states.

Viability-based resilience including management action:

Supposing that the future states of the system depend additionally on a management action policy $u(t)$, we get a new system, where the change of state of the system does not only depend on the starting point and the time, but also on the management action u taken, which is mathematically represented by

$$x'(t) = f(x, t, u). \tag{3.6}$$

Note that the original system is a special case of this system, where no action is taken at all, meaning that $u = 0$. With this additional opportunity, a system under disruption can possibly be supported by the right management action, while recovering. This can lead to a faster process of recovery, or can make the recovery process possible for states where recovery was not possible without management action. To adapt the concept of resilience, the definition of the viability kernel has to be extended in the following way.

$$Viab_{f,u}(A) = \{x \in A : \exists u : f(x, t, u) \in A \ \forall t \geq 0\}. \tag{3.7}$$

This means that there exists at least one management action policy, such that the system state stays in the set A for all positive times t. It is clear that this definition leads to a larger viability kernel since no management action is included by setting $u = 0$. Analogously, the definition of the capturing set can be extended, such that it contains all states, for which a management action exists, that drives the state of the system back to the target set within finite time. Applying the definition from above, the resilience set, or resilience basin, is the capturing set of the viability kernel of the set of desired states.

It includes all states that can be brought back to the viability kernel with an existing management action and be kept there with a suitable management action as well. Hence, if the state of the system after the disturbance lies within the resilience basin, a desired state can be restored. Note that this holds only if there is no new disturbance while the restoring process is running.

3.4 Conclusion

The paper summarizes and characterizes the different definitions and concepts of resilience. It is obvious that the concept is closely related to the understanding of stability of a dynamical system. In Pickl (1998), an algorithm is presented which judges if a certain set of matrices is stable or not.

The constructive algorithm represents a convex set by its extreme points and uses linear programming to construct the successive polytopes. Sufficient conditions for the finiteness of constructing the new polytope from the previous one, and for stopping the algorithm when either the set is proved stable or unstable are presented. Furthermore, a control-theoretic framework is developed which steers the dynamic system to a described equilibrium.

In a following contribution, this idea will be applied to the different applications presented and characterized in that article. Thereby, we are able to stabilize an environmental system, as well as a complex transportation system after a certain threat.

References

Adger, W. N. (2000): Social and ecological resilience, are they related? Prog. Hum. Geog. 24, 347–364.

Aitsi-Selmi, A. Egawa, S. Sasaki, H. Wannous, and C. Murray, V. (2015): The Sendai framework for disaster risk reduction: Renewing the Global Commitment to People's Resilience, Health, and Well-being. Int J Disaster Risk Sci 6 (2), 164–176. DOI: 10.1007/s13753-015-0050-9.

Alexander, D. E. (2013): Resilience and disaster risk reduction: an etymological journey. Nat. Hazards Earth Syst. Sci. 13 (11), 2707–2716. DOI: 10.5194/nhess-13-2707-2013.

Amorim, I. (2009): Resettlement of communities. The case study of Jaguaribara: A resilient Community (Northeast of Brazil). Jamba 2 (3), 216–234.

Aubin, Jean-Pierre (1990): A survey of viability theory. SIAM J. Control Optim. 28 (4), 749–788. DOI: 10.1137/0328044.

Aubin, J.-P. and Désilles, A. (2017): Traffic Networks as Information Systems. Berlin, Heidelberg: Springer Berlin Heidelberg, checked on 11/9/2018.

Bara, C. (2010). Naturgefahren & Ausfall von kritischen Infrastrukturen: Praxis-Forum vom 11. November 2010, Geoprotecta, St. Gallen. Unter Mitarbeit von Eidgenössische Technische Hochschule Center for Security Studies Zürich, IBCOL Technologies & Consulting AG Zweigniederlassung Küsnacht Sarnen und Schweiz Bundesamt für Bevülkerungsschutz.

Birkmann, J. (2008): Globaler Umweltwandel, Naturgefahren, Vulnerabilität und Katastrophenresilienz. Notwendigkeit der Perspektivenerweiterung in der Raumplanung. RuR (1), 2008.

Christmann, G. Ibert, O. Kilpert, H. Moss, and T. (2011): Vulnerabilität und Resilienz in sozio-räumlicher Perspektive. Begriffliche Klärungen und theoretischer Rahmen. Working Paper, Erker, Leibniz-Institut für Regionalentwicklung und Strukturplanung, 2011. In: www.irs-net.de/download/wp_vulnerabilitaet.pdf.

Cutter, S. L. (2016): Resilience to what? Resilience for whom? The Geographical Journal 182 (2), 110–113. DOI: 10.1111/geoj.12174.

Davoudi, S. Shaw, K. Haider, L. J. Quinlan, A. E. Peterson, G. D. Wilkinson, C. et al. (2012): Resilience: A Bridging Concept or a Dead End? "Reframing" Resilience: Challenges for Planning Theory and Practice Interacting Traps: Resilience Assessment of a Pasture Management System in Northern Afghanistan Urban Resilience: What Does it Mean in Planning Practice? Resilience as a Useful Concept for Climate Change Adaptation? The Politics of Resilience for Planning: A Cautionary Note. Planning Theory & Practice 13 (2), 299–333. DOI: 10.1080/14649357.2012.677124.

Detten, R. von Faber, F. and Bemmann, M. (Eds.) (2013): Unberechenbare Umwelt. Zum Umgang mit Unsicherheit und Nicht-Wissen. Wiesbaden: Springer Fachmedien Wiesbaden.

Gabriel, T. (2005): Resilienz - Kritik und Perspektiven. Zeitschrift für Pädagogik 51 (2005) 2, 207–217.

Gaillard, J.−C. (2007): Resilience of traditional societies in facing natural hazards. Disaster Prev and Management 16 (4), 522–544. DOI: 10.1108/09653560710817011.

Goecke, O. (2017): Risiko und Resilienz. Proceedings zum 11. FaRis & DAV Symposium am 9. Dezember 2016 in Köln (6).

Holling, C. S. (1973): Resilience and stability of ecological systems. Annual Review of Ecology and Systematics 1973 4 (1), 1–23.

Holling, C. S. Crawford, Stanley (1996): Engineering resilience versus ecological resilience. In: Engineering Within Ecological Constraints, Washington DC: National Academy Press, 31–43.

Holling C. S. (2001): Understanding the complexity of economic, ecological and social systems. Ecosystems 4, 390–405.

Holling, C. S. (2008): A definition of ecological resilience. http://www.geog.mcgill.ca/faculty/peterson/susfut/resilience/resilienceDef.html.

IPCC (2012): Annex II. Glossary of Terms. https://www.ipcc.ch/pdf/special-reports/srex/SREX-Annex_Glossary.pdf.

Jen, Erica (2003): Stable or robust? What's the difference? Complexity 8 (3), 12–18. DOI: 10.1002/cplx.10077.

Kelman, I. Gaillard, J. C. Lewis, James Mercer, and Jessica (2016): Learning from the history of disaster vulnerability and resilience research and practice for climate change. Nat Hazards 82 (S1), S. 129–143. DOI: 10.1007/s11069-016-2294-0.

Kharrazi, Ali (2018): Resilience. In: Encyclopedia of Ecology. Amsterdam: Elsevier, 414–418.

Lewis, J. (2013): Some realities of resilience: A case-study of Wittenberge. Disaster Prev and Management 22 (1), 48–62. DOI: 10.1108/09653561311301970.

Lewis, J. and Kelman, I. (2010): Places, people and perpetuity: Community capacities in ecologies of catastrophe. ACME: An International E-Journal for Critical Geographies (9), 191–220.

MacKinnon, D. and Derickson, K. D. (2013): From resilience to resourcefulness. In: Progress in Human Geography 37 (2), 253–270. DOI: 10.1177/0309132512454775.

Manyena, S. (2006): The concept of resilience revisited, Disasters, Vol. 30 No. 4, 433–450.

Manyena, S. B. O'Brien, G. O'Keefe, P. and Rose, J. (2011): Disaster resilience: a bounce back or bounce forward ability? In: Local Environment 16 (5), 417–424. DOI: 10.1080/13549839.2011.583049.

Martin, Sophie Deffuant, Guillaume Calabrese, and Justin M. (2011): Defining Resilience Mathematically: From Attractors To Viability. In Guillaume Deffuant, Nigel Gilbert (Eds.): Viability and Resilience of Complex Systems. Berlin, Heidelberg: Springer Berlin Heidelberg (Understanding Complex Systems), 15–36.

O'Hare, P. and White, I. (2013): Deconstructing resilience: Lessons from planning practice. Planning Practice & Research 28 (3), 275–279. DOI: 10.1080/02697459.2013.787721.

Pickl, S. (1998): Der τ-value als Kontrollparameter - Modellierung und Analyse eines Joint-Implementation Programmes mithilfe der dynamischen kooperativen Spieltheorie und der diskreten Optimierung, Aachen, Shaker Verlag, 193 Seiten.

Pugh, J. (2014): Resilience, complexity and post-liberalism. Area 46 (3), 313–319. DOI: 10.1111/area.12118.

Sudmeier-Rieux, K. (2014): Resilience—an emerging paradigm of danger or of hope? Disaster Prev and Management 23 (1), 67–80. DOI: 10.1108/DPM-12-2012-0143.

UNISDR (2017): Terminology. In: https://www.unisdr.org/we/inform/terminology

van der Vegt, G. S. Essens, P. Wahlström, M. and George, G. (2015): Managing risk and resilience. AMJ 58 (4), 971–980. DOI: 10.5465/amj.2015.4004.

Vugrin, E. D. Warren, D. E. and Ehlen, M. A. (2011): A resilience assessment framework for infrastructure and economic systems: Quantitative and qualitative resilience analysis of petrochemical supply chains to a hurricane. Proc. Safety Prog. 30 (3), 280–290. DOI: 10.1002/prs.10437.

Walker, B. Holling, C. S. Carpenter, S. R. and Kinzig, A. (2004): Resilience, adaptability and transformability in social-ecological systems. Ecology and Society 9 (2). Available online at www.jstor.org/stable/26267673, checked on 10/17/2018.

Wang, Z.; Nistor, M. S.; and Pickl, S. W. (2017): Analysis of the definitions of resilience. IFAC-PapersOnLine 50 (1), 10649–10657. DOI: 10.1016/j.ifacol.2017.08.1756.

Weichselgartner, J. Kelman, I. (2015): Geographies of resilience. Progress in Human Geography 39 (3), 249–267. DOI: 10.1177/0309132513518834.

White, I. O'Hare, P. (2014): From rhetoric to reality: Which resilience, why resilience, and whose resilience in spatial planning? Environ Plann C Gov Policy 32 (5), 934–950. DOI: 10.1068/c12117.

World Security Report 2018. September–October Issue 2018. http://www.worldsecurity-index.com/shareDir/documents/15409850700.pdf.

Yunes, M. A. (2003): Psicologia Positiva e Resiliencia: O foco no Individuo e na Fam'lia, Psicologia em Estudo, 8 (n. especial), 75–84.

4

Quantified Resilience Estimation of the Safety-Critical Traction Electric Drives

---

Igor Bolvashenkov[1,*], Hans-Georg Herzog[1] and Ilia Frenkel[2]

[1]Institute of Energy Conversion Technology, Technical University of Munich, Munich, Germany
[2]Center for Reliability and Risk Management, SCE – Shamoon College of Engineering, Beer-Sheva, Israel
E-mail: igor.bolvashenkov@tum.de; hg.herzog@tum.de;
iliaf@frenkel-online.com
*Corresponding Author

This chapter presents a new methodology of quantitative evaluation of the resiliency of electric traction drive which consists of the multilevel electric energy inverter and multiphase electric motor. It is proposed to consider such propulsion system as a system with a few states with degraded values of performance. As a complex indicator for resiliency estimation, it is suggested to take the criterion of degree of resiliency. For the approbation of the suggested method, the application case for assessment the resiliency of the traction train of an electrical helicopter was provided. The application results show that for design requirements on the value of reliability and resilience of electrical helicopter's propulsion system in operational conditions of the real flight only presented topology of multiphase electric motors in combination with multilevel electric inverters fully meet the project requirements without restriction. The presented methodology can be used successfully as a convenient tool for quantitative estimation and optimization of the degree of resiliency for safety-critical propulsion systems.

4.1 Introduction

Nowadays, with continuous increasing of complicacy and high integration of new engineering systems, the task of implementing the desired level of

87

Figure 4.1 Components of sustainable operation.

sustainable and resilience operations becomes very topical. Such a problem is closely linked to the task of very accurate evaluation of various operational sustainability indices of the whole propulsion system, shown in Figure 4.1. For the safety-critical system to determine correctly and to realize the desired value of resiliency is especially important.

Most important requirement for the vehicle traction electric drive is the value of resiliency. It means that the vehicle should have the possibility to operate sustainable in every degraded state after occurring several failures of electric traction drive components.

For practical realization of such requirements, all units and subsystem of traction drive should be resilient. As an application case of using the suggested technique, the resiliency value of two important units of the vehicle electric traction drive—the multiphase permanent magnet synchronous motor (PSM) of helicopter and electric energy inverter in conventional and multilevel versions were evaluated. In this instance, regarding to the particular requirements on the flight safety of helicopter, the complete failure probability for the whole propulsion system of helicopter should be less than 10^{-9}/h (Bolvashenkov et al., 2016).

Due to the constant need of traction electrical machines for special application, there are lot of studies which describe the results of qualitative analysis of various options of resilient traction drives components topologies for various vehicle applications, discussed in Bolvashenkov et al. (2016);

Table 4.1 Qualitative evaluation of resiliency (Bolvashenkov et al., 2016)

Phase Number Parameter	5 Phase	6 Phase	7 Phase	9 Phase
Overload capacity	8	9	9.5	10
Partial load mode	7	9	8	10
Torque ripple	7	8	10	9
Total	22	26	27.5	29

Bolvashenkov and Herzog (2016); Fazel et al. (2007); Fodorean et al. (2008); Josefsson et al. (2012); Sarrazin et al. (2011) and (Malinowski et al., 2010; Scuiller et al., 2010; Semail et al., 2008; Vigrianov, 2012; Villani et al., 2011). For example, in (Bolvashenkov et al., 2016) a technique for the qualitative resiliency assessment is suggested, which results of comparative analysis are shown in Table 4.1.

Current methodologies for assessment the value of resiliency of electrical machines, power electronics, and the computer components topologies are described in (Bavuso et al., 1987; De et al., 1998; Krivoi et al., 2006) and (Muellner and Thee, 2011; Trivedi, 2002; Ubar et al., 2008; Welchko et al., 2004). All these proposed techniques have similar disadvantage—the absence of universality. Each of these methods allows to solve a particular problem especially for a specific technical system.

The authors suggested a new universal approach and methodology of assessing the value of resiliency based on so-called degree of resiliency (*DOR*) of safety-critical technical systems as a whole, as well as the *DOR* of their units and subsystems.

4.2 Approach and Methodology

4.2.1 Degree of Resiliency

Taking into account the designation of the resiliency (*DOR*) of safety-critical technical system as a capacity to support the required level of the system performance in case of the failures occurring of its units, *DOR* can be determined as the time which the system can operate in degraded state without irreversible changes of performance. Mathematically, *DOR* can be represented by formula (4.1):

$$DOR_i = \frac{W_R}{W_N} \cdot \frac{\Delta t_i}{\Delta t_N}, \tag{4.1}$$

where W_R and W_N—different performance level, reduced and nominal, of safety-critical technical system; Δt_i, Δt_N—length of time of the system functioning after i-number of failures and in failure-free mode.

For different technical systems, as the term "performance" W, the productiveness, power, energy, quantity of information, can be regarded. The value of Δt_i is determined based on overload capacity of propulsion system after occurring i-number of failures.

In the case when the desired value of W_R is predefined by design requirements, it is advisable to evaluate *DOR* for each performance level based on formula (4.2):

$$DOR_{Ri}(W_R) = \frac{\Delta t_i}{\Delta t_N}. \tag{4.2}$$

The value of Δt_N is computed in depends of electric vehicles types and, accordingly, of the specific operating conditions of ships, trucks, aircraft, trains, buses, and cars. For instance, the electrical helicopter needs about 30 min for the safety providing of its function "search and rescue." Δt_i is estimated based on the overload capability and overheating thermal stability of safety-critical traction electric drive. The value of Δt_i is determined the time interval, when the safety-critical system can remain in operation after critical failures occurred without irreversible performance changes and impossibility of further operation.

The algorithm for *DOR* evaluation is described below:

- Analysis classification of all possible failures based on analytical methods;
- Comparison of all failures based on their criticality and importance;
- Assessment of possibilities and labor intensity of the failure recovery;
- Analysis of development and the possible effects of non-repairable failures;
- Determining the value of decreasing of the system's performance;
- Computation of an acceptable performance value regarding the project requirements;
- Calculation the ratio between values of demand and performance for each failure mode;
- When demand value exceeds the required value of performance, the accurate estimation of degraded state is carried out, and the value of Δt_i is computed;
- Based on above computed data, the *DOR* value can be calculated.

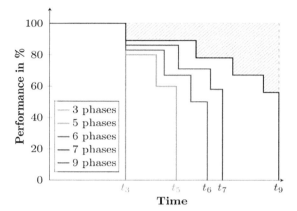

Figure 4.2 *DOR* of various multiphase motors.

In Figure 4.2, the values of *DOR* of various traction multiphase motor are represented graphically. Respectively, in Figure 4.3 the *DOR* values of conventional and multilevel power inverters are shown. The shaded area in Figure 4.2 the curve of the degradation of nine-phase electric motor is equated to the value of transition probability of this motor topology for the next degraded state.

The steps number in Figure 4.2 corresponds to the number of degraded states of the electric machine after safety-critical failure occurring until the moment when the motor of helicopter loses the performance completely. Considering the multiphase motor, each step corresponds to the next safety-critical open-phase failure. Thus, multiphase traction motor can be regarded as multistate system, which will be given special attention in Section 2.2.

Quantitative assessment of the total duration of the electric motor operations in all degraded states allows a quantitative comparative analysis of various possible topologies and design of the machine, considering their lifetime.

Figure 4.3 is a graphical representation of the qualitative comparison's results of the *DOR* of conventional and multilevel inverter for six-phase traction motors. The figure shows that *DOR* of the multilevel inverter is much superior to the conventional version.

In addition, the impact of noncritical failures on the resiliency value, which can lead to partial or temporal loss of system performance, can also be evaluated based on *DOR*, as shown in Figure 4.4.

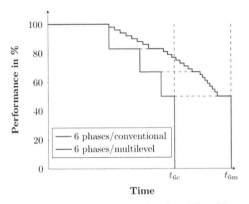

Figure 4.3   *DOR* of conventional and multilevel inverter.

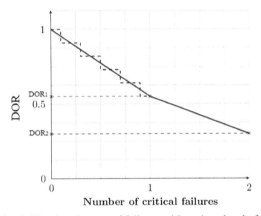

Figure 4.4   *DOR* values in case of failures with various level of criticality.

An original peculiarity of the suggested technique of quantified evaluation of *DOR* of safety-critical technical systems compared with modern methodologies is its universality, i.e., the opportunity of applying for various types of propulsion systems. As an application example, the evaluation of *DOR* and transition probabilities for Markov models of electrical helicopter traction drive is carried out.

4.2.2 Multistate System Reliability Markov Models and Transition Probabilities

Taking into account the design requirements on the total failure probability of electrical helicopter and statistical reliability data of traction electric machine

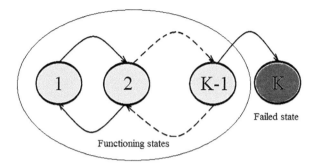

Figure 4.5 Multistate system state space diagram. (Krivoi et al., 2006)

and power electronics, it was concluded that the multistate system reliability Markov model (MSSR MM) is the optimal model for resiliency estimation of electrical helicopter traction drive. General view of the multistate system with K states is presented in Figure 4.5. The fundamental of this technique described in details in (Bolvashenkov and Herzog, 2015; Lisnianski et al., 2010; Natvig, 2011) and application cases in (Geyer and Schroder, 2010; Molaei, 2006; Ranjbar et al., 2011).

In the Markov model of the multistate system reliability with K states, the state number 1 corresponds to the failure-free operation of the system. The second and subsequent states (up to state K) correspond to degraded states. The final state K of MSSR MM corresponds to the total failure of propulsion system when the helicopter does not have the ability to fly.

Thus, the number of the degraded states of MSSR MM defined based on project requirements on the resiliency of electric traction drive.

The main task in constructing the MSSR MM is to calculate the number of degraded states and the values of transition probabilities. To define the values of transition probabilities λ_1, λ_2,...λ_K, the results of *DOR* evaluation for each state were applied:

$$\lambda_{Ri} = 1 - DOR_{Ri} \tag{4.3}$$

Here R is the degraded performance value regarding to design demands and i—number of safety-critical failures. The transition probability value strongly depends on a two groups of factors. They are the design factors and operational factors.

As can be seen from formula (4.3), to compute the values of transition probabilities for MSSR MM, the accurate computation of *DOR* is especially important. Application study of suggested technique to assess resiliency features of electric helicopter discussed in the following section.

4.3 Resilient Traction Drive

The electrical propulsion system of helicopter considering strict requirements on the reliability and resiliency is the safety-critical system.

4.3.1 Topology and Components

Generally, electric propulsion system of helicopter consists of four main components. They are energy storage, an electric converter, control unit, and traction motor, shown schematically in Figure 4.6.

In this study, the energy storage is not considered, but it should be investigated in the following researches.

4.3.2 Safety-Critical Failures

In research, report (Vigrianov, 2012) was concluded that all possible of failures in the system "electric converter-electric motor" can be diminished to four main types:

- short-circuit and open-circuit of motor phase;
- short-circuit and open-circuit of power electronics submodule.

In turn, both of failure type "short-circuit" can be diminished to failure type "open-circuit" based on special design of the motor stator winding performance, using of modern insulation materials and advanced manufacture technology.

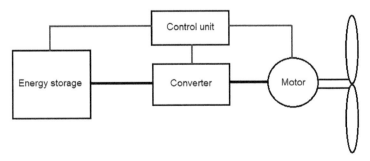

Figure 4.6 Components of electric traction drive.

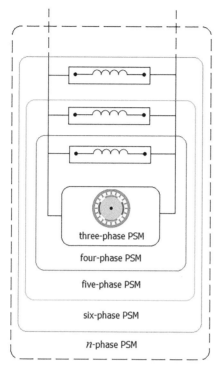

Figure 4.7 Schematic presentation for a multiphase motor.

4.3.3 Multiphase Electric Motor

Preliminary studies (Bolvashenkov et al., 2016) shown that conventional three-phase electric motor cannot meet the stringent requirements on the resiliency of the whole electric traction drive of designed electrical helicopter.

One known possibility to realize the resilient traction electric motor is to improve the number of motor phases. In Figure 4.7, the multiphase electric motor as a multistate system schematically represented.

Such approach allows to reduce the phase currents and to implement the converter submodules integrally manufactured. In addition, it increases the efficiency of the electric motor and decreases the torque ripples.

Based on the diversity of various schemes, connection methods of power supplies and inverters as well as switching algorithms of the stator phases of electric motor, it is possible to implement the system using automatically changed structures and parameters, respectively, depending on the purpose and functioning conditions.

Based on the failure probability values of one phase of electrical machine the optimal number of phases for the given resiliency level can be determined.

In paper (Bolvashenkov et al., 2016) is concluded that optimal electric machine for safety-critical application regarding system approach, is a multiphase PSM with concentrated stator windings and internal v-shaped arrangement of permanent magnets on the rotor.

Taking into account a normal, failure-free operational mode, the electric motor can tolerate a short-term overload based on large thermal capacity. At the same time for failure cases the situation is changing significantly. Operational experience shows (Hann, 2010; Katzman, 2001; Mahdavi et al., 2013) that the most operational failures are caused by technological overloads an overheating.

The most common types of failures which can lead to the critical overheat of the windings insulation and to a complete failure of traction electric machine and whole drive, are:

- the short circuit between turns;
- the short circuit between coils;
- the short circuit between motor phases;
- the short circuit between wires and the motor housing.

To maintain sustainable the performance of safety-critical traction drive at a required level, the phase currents in remaining phases of the electric machine should be increased. Due to overcurrent and overheating, the operational time in such degraded load mode should be limited.

Considering electric propulsion system of helicopter such load levels in degraded operation are: for long-time operation—65% of the rated value, and for short-time operation—85% and 113%.

To evaluate the overheating value of the windings insulation of multiphase electric machines, it is suggested to consider the overheating factor K_T, which indicates the degree of exceeding the rated value of windings insulation temperature:

$$K_T = \frac{T_i}{T_N} \tag{4.4}$$

where T_i and T_N are windings insulation temperature in failed mode in i-phases and in the failure-free mode, respectively. Factor K_T is shown in Figure 4.8.

Table 4.2 demonstrates the results of a preliminary evaluation of the overload capacity of multiphase motors in various load modes and the open-circuit failure of one or two phases.

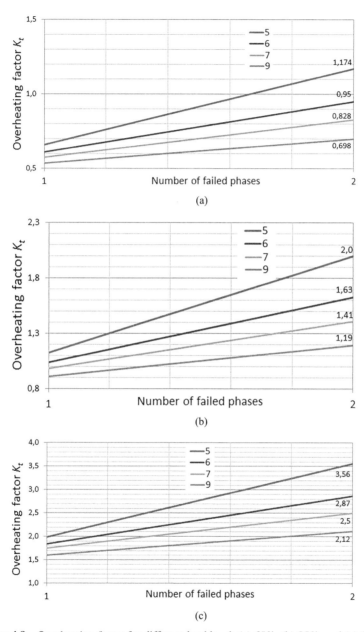

Figure 4.8 Overheating factor for different load level: (a) 65%, (b) 85%, and (c) 113%.

Table 4.2   Comparison of overload capacity

Phase number	5		6		7		9	
Fault number	1	2	1	2	1	2	1	2
Load level								
100%	0	0	0	0	0	0	+	0
85%	+	0	+	0	++	0	++	+
75%	++	0	++	+	++	+	++	++
65%	++	+	++	++	++	++	++	++
50%	++	++	++	++	++	++	++	++

Table 4.2 has the following designations, which allow to ealuate the over-load magnitude: **++**—no overload; **+**—less than 15%; **0**—more than 15%.

Thus, the basic idea of the preliminary overload analysis is to find the critical operational points considering overload and overheating. In operational mode, when overcurrent exceeds the above given values, this mode is regarded as safety-critical.

Regarding the low overload capacity rates and overheating stability of five-phase electric machine, as shown in Figure 4.8, this option was eliminated from the following assessment.

In failure cases, the negative effects of operational overloads are an overcurrent and overheating of the motor, which can lead to decreasing of the motor lifetime, as shown in Figure 4.9.

Basic characteristics for *DOR* computation are the thermal curves of electric drive components, estimated by formulas Katzman (2001):

$$t_N = \frac{\ln K^2 - \ln(K^2 - 1)}{A/C}, \qquad (4.5)$$

where t_N—is the time to reach the rated value of the motor temperature, K—the rate of the phase current exceeding, A—the motor thermal radiation, and C—the heat capacity of traction motor.

In Figure 4.10, the results of thermal analysis of traction motor behavior in overload modes are presented. The values of N_{deg} and N_{nom} are the values of traction electric drive performance in degraded and nominal modes.

Based on the thermal analysis of the motor, the Δt_i value for various options of electric machine and the different failure cases can be computed. Considering the value Δt_i, the transition probabilities for MSSR MM were determined.

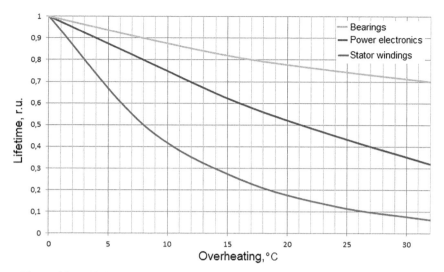

Figure 4.9 Lifetime of electric drive components (Bolvashenkov and Herzog, 2015).

Dependency graphs of *DOR* at 65%, 85%, and 113% of rated load, on the number of safety-critical failures are presented in Figure 4.11(a)–(c), respectively.

In common case, these graphs allow to provide the comparative analysis of various options of power drive with different quantity of safety-critical faults. This can be very useful in the field of design and development of advance electric propulsion systems, considering the requirements on the resiliency and safety.

On the basis of above-mentioned graphs, a comparative *DOR* estimation for selected options of electric machines can be provided. Nevertheless, more informative and suitable for practical applications, according to the authors, are the dependences of *DOR* on the value of required load level, shown in Figure 4.12. In this case, it is not difficult to evaluate the accordance of resiliency features for each compared variant to project requirements.

4.3.4 Electric Inverter

Regarding the current developments in the field of power electronics, for the comparative analysis of the resiliency has been considered two options of electrical power inverter: conventional and multilevel, which are presented in Figure 4.13. Well-known voltage source inverter is the most commonly

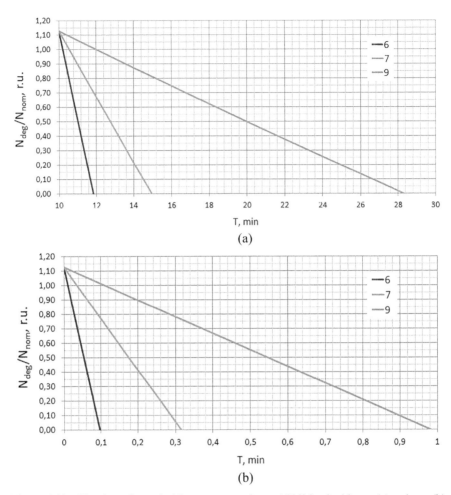

Figure 4.10   The time of sustainable motor operation at 113% load with one (a) and two (b) safety-critical failures.

used electric converter between the series-parallel-configured batteries and the traction motor in electric propulsion systems.

As a conventional topology (Figure 4.13a), the scheme of the six-pulse-bridge (B6-bridge) has been considered. That means in this case one single inverter for multiple phases drives the multiphase synchronous motor.

For comparative analysis, as competitive option of power electronics of the multilevel cascaded H-bridge (CHB) inverter (Figure 4.13b) has been

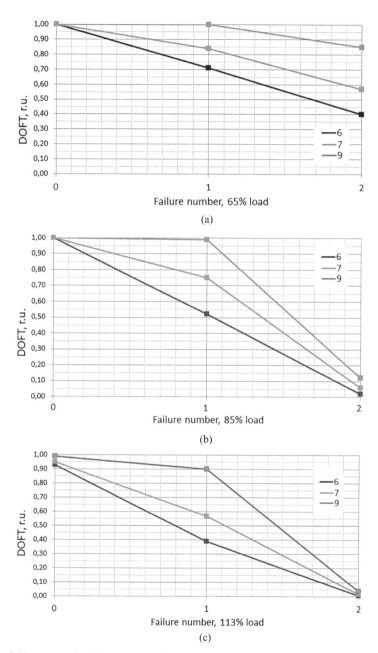

Figure 4.11 *DOR* of multiphase motor in failure modes: (a) 65%, (b) 85%, and (c) 113% of rated load.

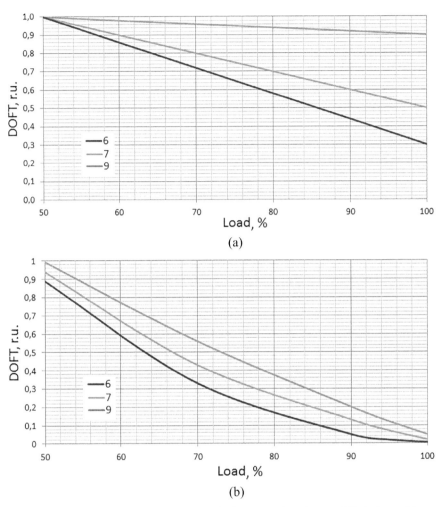

Figure 4.12   *DOR* for the given load level in failure modes: (a) one failed phase, (b) two failed phases.

discussed. By this topology each phase of electric motor is connected through own CHB inverter with own energy storage.

The multilevel inverters have well-known benefits, discussed in Malinowski et al., (2010). They are the high modularity, small filters, low power losses, and high fault tolerance. Taking into account the results of (Fazel et al., 2007; Josefsson et al., 2012; Malinowski et al., 2010; Sarrazin et al., 2011),

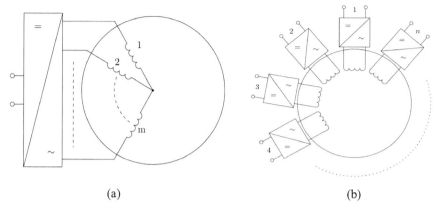

Figure 4.13   Electric inverter topologies: (a) conventional, (b) multilevel.

CHB inverter needs the lowest number of components compared with other multilevel inverter topologies.

From the point of view of reliability and fault tolerance, CHB multilevel inverter has also an important advantage for the electrical traction drive because of its opportunity to operate with unbalanced or faulty direct current electric energy sources. This feature has a positive impact on the life time of battery of electric vehicle.

Considering the project requirements for helicopter propulsion system in terms of resiliency, the best topology of multilevel CHB inverter is a 17-Level inverter topology. Thus, the number of submodules included in one phase of traction electric motor is eight units, as shown in Figure 4.14. As a basic semiconductor component, the MOSFETs were selected. By using of MOSFETs in multilevel CHB inverter, by reason of the low level of electric losses it is possible to improve the efficiency of the whole inverter. Either energy storage module is bound to the H-bridge with four MOSFETs.

Figure 4.14 represents the electrical scheme of one inverter submodule with four MOSFETs (a), and its general view and total weight (b).

During operation in overload modes the temperature of semiconductors increases significantly. This happens for a reason of the low heat capacity of MOSFET. In the worst case when the temperature of MOSFET exceeds a critical value, the semiconductor component fails. In failures cases by reason of overload modes the current and temperature of inverter increase drastically. This reduces reliability and resiliency of the whole traction drive and significantly decreases their lifetime. Based on statistical reliability data,

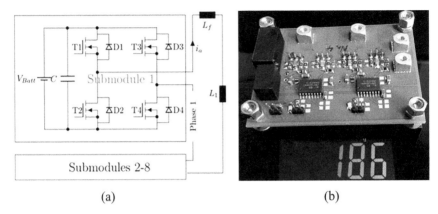

(a)   (b)

Figure 4.14   Submodule of 17-level inverter: (a) electrical scheme, (b) general view and weight.

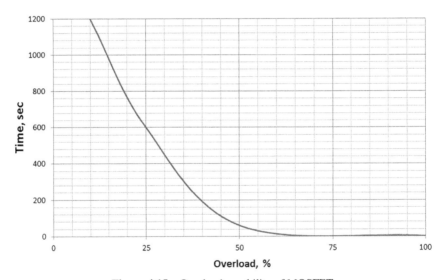

Figure 4.15   Overload capability of MOSFET.

the temperature is the main factor to estimate the overload capacity of power electronic devices in the failure cases.

In Figure 4.15, the schedule is shown, which is built based on the experimental data considering the normative overload capacity of MOSFETs.

An experimental study on the laboratory test bench has confirmed the validity of the use of these dependencies for the calculation the value of *DOR*.

Table 4.3 Overload of conventional inverter in failure modes

Phase number	6			7			9		
Fault number	1	2	3	1	2	3	1	2	3
Load level									
113%	+	MF	MF	++	+	MF	+++	+++	++
85%	+++	+	MF	+++	+++	+	+++	+++	+++
65%	+++	+++	MF	+++	+++	+++	+++	+++	+++

Table 4.4 Overload of multilevel inverter in failure modes

Phase number	6			7			9		
Fault number	1	2	3	1	2	3	1	2	3
Load level									
113%	+	PhF	PhF	++	+	PhF	+++	+++	++
85%	+++	++	+	+++	+++	++	+++	+++	+++
65%	+++	+++	+++	+++	+++	+++	+++	+++	+++

On the basis of thermal analysis of power electronics in failure cases, critical operational point considering the thermal stability of semiconductor devices have been identified as shown in Tables 4.3 and 4.4.

Tables 4.3 and 4.4 have the following designations: + + +—overload is less than 10%; ++—overload is less than 25%; +—overload is less than 40%; MF—motor failure; PhF—phase failure.

Qualitative analysis of the data presented in the tables shows that the 17-level inverter significantly exceeds the level of resiliency of the inverter with a conventional topology.

4.4 Results of Simulation

For resiliency evaluation of the traction electric motor and electric inverter and for their compliance estimation with project requirements, the MSSR MM, described in details in Lisnianski et al. (2010) and Natvig, (2011), was built.

To create the models of the multiphase traction electric motor and CHB inverter, the traction electric drive was regarded as a system with redundant resiliency, i.e., the system with loaded functional redundancy.

Taking into account the requirements on resiliency of safety-critical drives and failures statistics of the electric motor power inverter from

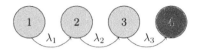

Figure 4.16   MSSR MM of electric motor.

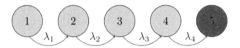

Figure 4.17   MSSR MM of electric inverter.

Table 4.5   Probabilities of total PSM failure

Phase number	6	7	9
65% nominal load	$2.85 \cdot 10^{-11}$	$2.38 \cdot 10^{-12}$	$6.80 \cdot 10^{-16}$
85% nominal load	$8.19 \cdot 10^{-9}$	$1.27 \cdot 10^{-10}$	$5.42 \cdot 10^{-12}$
113% nominal load	$6.00 \cdot 10^{-6}$	$2.71 \cdot 10^{-6}$	$6.10 \cdot 10^{-10}$

(Bolvashenkov et al., 2016; Lauger, 1982; Ermolin and Zerichin, 1981; Geyer and Schroder, 2010), the MSSR MM structures with number of degraded states for total failure probability analysis was determined. The state-space diagrams of MSSR MMs for traction motor and electric power inverter shown, respectively, in Figures 4.16 and 4.17.

The number of states of the Markov model is determined by the probability of failure-free operation of traction electric drive and resiliency requirements established by the design documentation.

In both above figures, the first states correspond to a failure-free operational mode of components: motor and inverter. The other grey states correspond to several degraded modes. In case of motor model (Figure 4.16) they are two states and in case of inverter model (Figure 4.17) they are three states.

The red states of diagrams correspond to the total failed components of electric propulsion system when helicopter is not able to safety fly operation.

The total failure probabilities of multiphase traction motors with 6, 7, and 9 phases at the 113% load level are shown in Table 4.5 and in Figure 4.18.

The results of simulation allow to quantify the degree of resiliency of multiphase motors with various phase number. The nine-phase topology of electric motor is most promised considering project requirements on the resiliency of vehicle traction drives.

The obtained results confirm the results of the studies presented in Semail et al. (2008) and Vigrianov (2012), that seven-phase electric traction motors can be operated after the loss of two phases during a limited time, at nominal load (limitation because of thermal stability), and for a long time with a reduced load.

Considering the project requirements on the resilience, the reliability features of electric inverter was investigated based on the Markov approach. The corresponding graphs for various number phases number and two inverter topologies are shown in Figures 4.19 and 4.20.

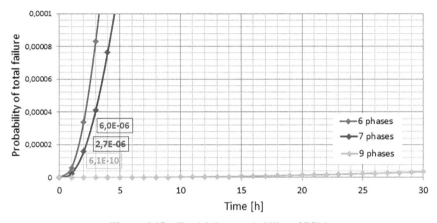

Figure 4.18   Total failure probability of PSM.

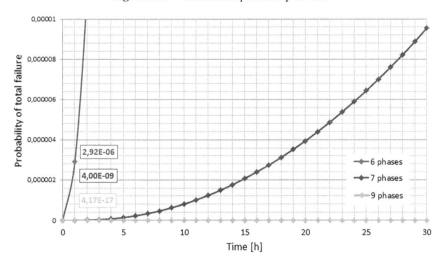

Figure 4.19   Total failure probability of motor with conventional inverter.

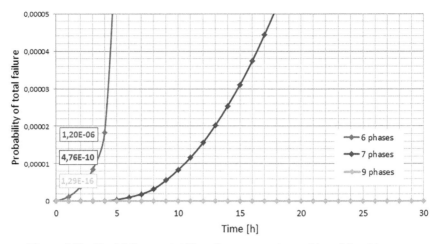

Figure 4.20   Total failure probability of one motor phase with multilevel inverter.

Table 4.6   Probabilities of total inverter failure

Phase number	6	7	9
Topology B6 (Motor total failure)	$2.9 \cdot 10^{-6}$	$4.0 \cdot 10^{-9}$	$4.2 \cdot 10^{-17}$
CHB (Phase total failure)	$1.2 \cdot 10^{-6}$	$4.8 \cdot 10^{-10}$	$1.3 \cdot 10^{-16}$

The results of the probability calculations of a failure-free operation of the electric inverter during one operational hour at 113% of nominal load for two inverter's topologies are shown in Table 4.6.

The results of simulation allow to quantify the degree of resiliency of two inverter topologies, conventional and multilevel inverter. In this case, seven- and nine-phase topologies of electric motor are most promised considering project requirements on the resiliency of vehicle traction drives.

4.5 Conclusion

The chapter presents a new approach and technique for evaluating the resiliency of a safety-critical electric traction drive. The presented assessment of resilient topologies of electric traction drives is well formalized and suitable for practical application in reliability engineering. This methodology can be used successfully to determine the quantitative value of resiliency of the main components of electric traction drive, such as multiphase traction motors and multilevel electric inverter, taking into account the aging process of components and impact of real operational conditions.

More broadly, the suggested approach and methodology can be implemented as a universal tool to define the degree of resiliency of safety-critical technical systems, and to optimize its value considering all possibilities, such as redundancy, monitoring, predictive control, and various maintenance strategies.

References

Bavuso, S. J., Dugan, J. B., Trivedi, K. S., Rothmann, E. M. and Smith, W. E. (1987). "Analysis of Typical Fault-Tolerant Architectures using HARP," in IEEE Transactions on Reliability, Vol. R-36, Iss. 2, pp. 176–185.

Bolvashenkov, I., Kammermann, J., Willerich, S. and Herzog, H.-G, (2016). "Comparative Study of Reliability and Fault Tolerance of Multi-Phase Permanent Magnet Synchronous Motors for Safety-Critical Drive Trains," in Proceedings of the International Conference on Renewable Energies and Power Quality (ICREPQ'16), 4th to 6th May, Madrid, Spain, pp. 1–6.

Bolvashenkov, I. and Herzog, H.-G. (2016). "Use of Stochastic Models for Operational Efficiency Analysis of Multi Power Source Traction Drives," in Proc. of IEEE of International Symposium on Stochastic Models in Reliability Engineering, Life Science and Operations Management (SMRLO'16), 15–18, Beer Sheva, Israel, 2016, pp. 124–130.

Bolvashenkov, I. and Herzog, H.-G. (2015). "Approach to Predictive Evaluation of the Reliability of Electric Drive Train Based on a Stochastic Model," in Proc. of IEEE 5th International Conference on Clean Electric Power (ICCEP'15), 16–18 June 2015, Taormina, Italy, pp. 1–7.

Brazhnikov, A. V. and Belozyorov, I. R. (2011). "Prospects for Use of Multiphase Phase-Pole-Controlled AC Inverter Drives in Traction Systems," European Journal of Natural History, Russia, Vol. 2, pp. 47–49.

De Pra, U., Baert, D. and Kuyken, H.(1998). "Analysis of the Degree of Reliability of a Redundant Modular Inverter Structure," in Proc. of IEEE 12th International Telecommunications Energy Conference, 04-08 Oct. 1998, San Francisco, CA, pp. 543–548.

Ermolin, N. P. and Zerichin, I. P. (1981). "Zuverlssigkeit elektrischer Maschinen," Berlin, Verlag Technik, p. 227. (in German).

Fazel, S., Bernet, S., Krug, D. and Jalili, K. (2007). "Design and Comparison of 4-kV Neutral-Point-Clamped, Flying-Capacitor, and Series-Connected H-Bridge Multilevel Converters," IEEE Transactions on Industry Applications, Vol. 43, No. 4, Jul.–Aug. 2007, pp. 1032–1040.

Fodorean, D., Ruba, M., Szabo, L. and Miraoui, A. (2008). "Comparison of the Main Types of Fault-Tolerant Electrical Drives used in Vehicle Applications," In Proc. of International Symposium on Power Electronics, Electrical Drives, Automation and Motion, (SPEEDAM), June 11–13, Ischia, Italy, pp. 895–900.

Geyer, T. and Schroder, S. (2010). "Reliability Considerations and Fault-Handling Strategies for Multi-MW Modular Drive Systems," in IEEE Transactions on Industry Applications, Vol. 46, No. 6, pp. 2442–2451.

Hann, D. (2010). "A Combined Electromagnetic and Thermal Optimisation of An Aerospace Electric Motor," in Int. Conference on Electrical Machines, ICEM, 6-8 Sept. 2010, Rome, Italy, pp. 1–6.

Josefsson, O., Thiringer, T., Lundmark, S. and Zelaya, H. (2012). "Evaluation and Comparison of a Two-level and a Multilevel Inverter for an EV using a Modulized Battery Topology," in Proc.of IEEE 38th Annual Conference on Industrial Electronics Society (IECON), Oct. 25–28, Montreal, Canada, pp. 2949–2956.

Katzman, M. M. (2001). "Electrical machines," Akademia, Moscow, Russia, p. 463. (in Russian).

Krivoi, S., Hajder, M., Dymora, P. and Mazurek, M. (2006). "The Matrix Method of Determining the Fault Tolerance Degree of a Computer Network Topology," Sofia, Bulgaria, Publisher: ITHEA, Vol. 13, No. 3, pp. 221–227.

Lauger, E. (1982). "Reliability in Electrical and Electronic Components and Systems," North - Holland Publ. Co., Amsterdam, p. 1171.

Levi, E. (2008) "Multiphase Electric Machines for Variable-Speed Applications," IEEE Transactions on Industrial Electronics, Vol. 55, No. 5, pp. 1893–1909.

Lisnianski, A., Frenkel, I. and Ding, Y. (2010). "Multi-State System Reliability Analysis and Optimization for Engineers and Industrial Managers," Berlin, New York, Springer, p. 393.

Mahdavi, S., Herold, T. and Hameyer, K. (2013). "Thermal Modeling as A Tool to Determine the Overload Capability of Electrical Machines," International Conference on Electrical Machines and Systems (ICEMS), 26–29, Busan, Korea, 2013, pp. 454–458.

Malinowski, M., Gopakumar, K., Rodriguez, J. and Perez, M.(2010). "A Survey on Cascaded Multilevel Inverters," IEEE Transaction on Industrial Electronics, Vol. 57, No. 7, pp. 2197–2206.

Muellner, N. and Thee, O. (2011). "The Degree of Masking Fault Tolerance vs. Temporal Redundancy," in IEEE Workshops of International Conference on Advanced Information Networking and Applications (WAINA), 22–25 March 2011, Biopolis, Singapore, pp. 21–28.

Natvig, B. (2011). "Multi-state Systems Reliability Theory with Applications," John Wiley & Sons, New York, p. 232.

Ranjbar, A. H., Kiani, M. and Fahimi, B. (2011). "Dynamic Markov Model for Reliability Evaluation of Power Electronic Systems," in Proc. of IEEE International Conference on Power Engineering, Energy and Electrical Drives (POWERENG), Malaga, Spain, pp. 1–6.

Sarrazin, B., Rouger, N., Ferrieux, J. P. and Crebier, J. C. (2011). "Cascaded Inverters for Electric Vehicles: Towards a Better Management of Traction Chain from the Battery to the Motor?", in Proc. of IEEE International Symposium on Industrial Electronics, June 27–30, Gdansk, Poland, pp. 153–158.

Scuiller, F., Charpentier, J.-F. and Semail, E. (2010). "Multi-Star Multi-Phase Winding for a High Power Naval Propulsion Machine with Low Ripple Torques and High Fault Tolerant Ability," in Proc. of the IEEE Vehicle Power and Propulsion Conference (VPPC), 1–3 Sept. 2010, Lille, France, pp. 1–5.

Semail, E., Kestelyn, X., and Locment, F. (2008). "Fault Tolerant Multiphase Electrical Drives: the Impact of Design," European Physical Journal - Applied Physics (EPJAP), Vol. 43, Iss. 2, pp. 159–162.

Trivedi, K. S. (2002). "Probability and Statistics with Reliability, Queuing, and Computer Science Applications," Second edition, Wiley, p. 848.

Ubar, R., Devadze, S., Jenihhin, M., Raik, J., Jervan, G. and Ellervee, P.(2008). "Hierarchical Calculation of Malicious Faults for Evaluating the Fault-Tolerance," in Proc. of 4th IEEE International Symposium on Electronic Design, Test & Applications (DELTA), 23–25 Jan. 2008, Hong Kong, pp. 222–227.

Vigrianov, P. G.(2012). "Assessment the Impact of Different Failures on the Power Characteristics of the Low Power 7-phase Permanent Magnet Synchronous Motors," Moscow, Journal "Questions to Electromechanics," Moscow, Vol. 128, Iss. 3, pp. 3–7. (in Russian).

Villani, M., Tursini, M., Fabri, G., and Castellini, L. (2011). "Multi-Phase Permanent Magnet Motor Drives for Fault-Tolerant Applications," in Proc.

of IEEE International Electric Machines & Drives Conference (IEMDC), 5–18 May 2011, Niagara Falls, Canada, pp. 1351–1356.

Welchko, B. A., Lipo, T. A., Jahns, T. M. and Schulz, S. E. (2004). "Fault Tolerant Three-Phase AC Motor Drive Topologies: A Comparison of Features, Cost, and Limitations," IEEE Transactions on Power Electronics, Vol. 19, No. 4, pp. 1108–1116.

5

Bayes Decision-Making Systems for Quantitative Assessment of Hydrological Climate-Related Risk using Satellite Data

Yuriy V. Kostyuchenko[*], Maxim Yuschenko, Ivan Kopachevsky and Igor Artemenko

Scientific Centre for Aerospace Research of the Earth, National Academy of Sciences of Ukraine
E-mail: yuriy.v.kostyuchenko@gmail.com; max.v.yuschenko@gmail.com; ivankm@ukr.net; igor.artemenko@casre.kiev.ua
*Corresponding Author

The task of constructing the method of assessment of the climate-related hydrological risks based on the satellite observation and ground measurement data, and its application both in the regional (north-western part of Ukraine) and in the local (basins of the Stokhid and Prypyat rivers) levels are described in this chapter. The model of expansion of dangerous processes is proposed; basing of the proposed model the method of processing of satellite data and analysis of the spectral indicators is constructed. The satellite observation data from Landsat satellites and field data for the period 1975–2015 in the study region was collected and processed. The regional distributions of climatic indicators over the observation period have been analyzed; key factors influencing the state of hydrological safety, as well as ecosystem climate-related changes and reactions have been identified. An analysis of the ecosystem reaction based on the study of spectral response was conducted. The changes in the spectral indices that can be used as informative signs for assessing the risks of climate-related hydrological processes, revealing through plant heat and water stresses are determined. The Bayes procedure for risk assessment was developed on this base. To determine the probability of manifestation of stress in a set of spectral characteristics, an equation is proposed, parameters of key variables are defined. The method of assessing the complex risk associated with hydrological threats is proposed. The risks of flooding on the study territory are mapped.

5.1 Introduction

Implementation of quantitative assessments of hazardous phenomena using remote-sensing data is wide-spread practice and important task, taking into account the increasing number and intensity of catastrophes in the world over the past decades (Guha-Sapir et al., 2011), including climate-related cases (Climate Change, 2001). Whereas the most dangerous and drastic changes are expecting in polar and tropic regions, temperate zone should also be analyzed from viewpoint of vulnerability toward climate change, as well as the climate-related hazards risks should be assessed. In particular, variations of hydrological processes and flood risks escalations, connected with climate changes, threatening the ecology, infrastructure, population, agriculture, and industry. So, the region on the border of Ukraine, Poland, and Byelorussia, with fragile ecosystems, vulnerable energy infrastructure, agriculture, and complicated water management system, might be interesting case for risk analysis. For this region, the task of climate-related risks assessment is extremely urgent.

Utilization of satellite data for construction of technique of quantitative risks assessment may support the necessary quality of decision making (CEOS/NOAA, 2001; Regulation (EU) 2010). Number of applied methods were developed during last decades (Bartell et al., 1992; Ermoliev and Hordijk, 2006; Fisher and Woodmansee, 1994), directed to calculation of quantitative assessments of the risks, generated by natural and technological threats. Bayesian methods are a simple and effective tool for evaluating probabilistic measures of various risks (Fenton and Neil, 2012). However, in this case, it is necessary to correctly estimate the distribution of initial data, for which it is necessary to apply complex models of the systems under study (Jiang and Mahadevan, 2007). This chapter is purposed to demonstrate the possibility to construct a specific methodology to utilize the satellite and ground measurement data for assessment of hydrological climate-related risks.

5.2 Methodological Notes

5.2.1 Generalized Stochastic Model of Hydrological Threats

Method of hydrological risks assessment using satellite data should be based on the model of dangerous processes. Hazardous dynamics (floods and swamping) might be described as the variation of moisture and water content in the accumulation zone s1 and in the discharge zone s_2, according to (Rodriguez-Iturbe et al., 1991):

$$ds_1 = A(s_1, s_2)dt + B(s_1, s_2)dW_t, \tag{5.1}$$

$$ds_2 = C(s_1, s_2)dt + D(s_1, s_2)dW_t. \tag{5.2}$$

Here dW_t – is the Wiener increment: $\langle dW_t \rangle = 0$, $\langle dW_t dW_{t'} \rangle = 1$ if $t = t'$, and $dW_t = 0$ for all any cases. This increment uses for description of long-term fluctuations of evapotranspiration and precipitation, or describes the parameter:

$$\alpha = LE_p/2wu, \tag{5.3}$$

where L – is the total size of studied area, E_p – evapotraspiration, w – average volume of atmospheric humidity, u – average wind speed.

Functions $A(s_i)$, $B(s_i)$, $C(s_i)$ i $D(s_i)$ may be determined as (Rodriguez-Iturbe et al., 1991):

$$A(s_1, s_2) = \frac{P_a}{nz_r}\{1 + \langle\alpha\rangle[(1 - f_g)s_1^c + f_g s_2^c]\}(1 - \varepsilon s_1^r) - \frac{E_p s_1^c - k_s s_1^b}{nz_r}, \tag{5.4}$$

$$B(s_1, s_2) = \frac{P_a}{nz_r}[(1 - f_g)s_1^c + f_g s_2^c](1 - \varepsilon s_1^r)\sigma, \tag{5.5}$$

$$C(s_1, s_2) = \frac{P_a}{nz_r}\{1 + \langle\alpha\rangle[(1 - f_g)s_1^c + f_g s_2^c]\}(1 - \varepsilon s_2^r) - \frac{E_p s_2^c - Q_2(t)}{nz_r}, \tag{5.6}$$

$$D(s_1, s_2) = \frac{P_a}{nz_r}[(1 - f_g)s_1^c + f_g s_2^c](1 - \varepsilon s_2^r)\sigma. \tag{5.7}$$

Here P_a – precipitation; f_g – partition of territory, covered by discharge zone; $\varepsilon, r, c,$ – empirical coefficients; n – porosity; z_r – thickness of the active root layer; and the flows of underground water Q_i may be described as (Entekhabi et al., 1992):

$$Q_1(s_1) = k_s s_1^b, \tag{5.8}$$

$$Q_2(t) = \left(\frac{1 - f_g}{f_g}\right)\frac{k_s}{J}\int_{-\infty}^{t} s_1^b(t - \tau)e^{-\tau/J} d\tau, \tag{5.9}$$

where k_s – permeability with the total saturation, t, τ – time, b – empirical coefficient, J – average groundwater delay, describing as (Krayenhoff van de Leur, 1958) $J = S_y l^2/\pi^2 T$, where S_y – debit of saturated zone, T – average permeability, l – average distance between discharge zones (drainage density). Parameters J and f_g describing the hydrogeological (parameters of water table) and geomorphological (location and size of discharge zones) features of the surface.

Combined Penman-Monteith equation could be used to calculate average daily evapotranspiration on the level on the top of vegetation (Budyko, 1974):

$$
E_p = \left\{ \frac{[f(A)+1]\{R_n - G\}\Delta}{[\sigma f(A)+1]C_p\rho} + [f(A)+1]\frac{\rho_2^* - \rho_2}{r_a} \right\}
$$
$$
\times \left\{ \frac{r_a + r_x}{r_a} + \frac{[f(A)+1]L_v\Delta}{[\sigma f(A)+1]C_p\rho} \right\}^{-1}. \tag{5.10}
$$

Here Δ – derivative of saturated vapor pressure; C_p – specific heat of air at the constant pressure; L_v – latent heat of transformation of water into vapor; σ – the ratio of the area of convection to the area of evapotranspiration; r_a – resistance from the atmosphere to vapor motion from the vegetation surface; R_n – amount of solar heat entering to the surface of evapotranspiration; G – the amount of energy that goes from vegetation to the soil for a certain time; r_x – resistance from the surface of the evaporator to the exit of water vapor; ρ – air density calculated with average pressure and actual humidity; ρ_2^* – vapor density with full saturation at daily average temperature; ρ_2 – actual vapor density in the atmosphere over the vegetation cover; *f(A)* – effective area of vegetation per unit of the total area of the investigated territory.

Thus, the task of control of hydrological and hydrogeological risks (in particular, flooding and swamping) using satellite data may be reduced to determination of the methodology for analyzing the set of surface indicators corresponding to the variable of Equations (5.1–5.10), and thus to control of changes of indicators of the reaction of local ecosystems to changes of the water and heat balance according to certain types of terrestrial cover.

5.2.2 Spectral Model of Surface Response to Heat and Water Stress

The surface reflection spectra, detected using satellite sensors, are forming by integrating energy input from large areas. This essentially distinguishes them from the spectra obtained by ground measurements, and should be taken into account in the comparative analysis. In some narrow spectral bands, these effects may not be essential. However, when the bandwidth is considerable, or when the composition of spectral bands (spectral indices) should be analyzed, the spatial and temporal variability of the energy fluxes from the surface should be taken into account.

The energy balance of surface site R_n may be described as follow (Gupta et al., 1999; Peters-Lidard et al., 1997):

$$R_n(T_{aero}, \mathbf{P}, \mathbf{W}) = H(T_{aero}, \mathbf{P}, \mathbf{W}) + LH(T_{aero}, \mathbf{P}, \mathbf{W}, \mathbf{\Theta})$$
$$+ G(T_{aero}, \mathbf{P}, \mathbf{W}, \mathbf{T}, \mathbf{\Theta}). \qquad (5.11)$$

Here T_{aero} – aerodynamic surface temperature, LH – latent heat (the residual heat energy that came with radiation after absorption by vegetation and soil); parametric description of soils and vegetation ($\mathbf{P}$), precipitation and solar radiation ($\mathbf{W}$), thermal ($\mathbf{T}$), and hydrological ($\mathbf{\Theta}$) parameters of surface.

Variations of thermal and hydrological parameters may be described as (Castelli et al., 1999):

$$\frac{d\mathbf{T}}{dt} = f(T_{aero}, \mathbf{\Theta}, \mathbf{T}, \mathbf{P}, \mathbf{W}), \qquad (5.12)$$

$$\frac{d\mathbf{\Theta}}{dt} = f(LH, \mathbf{\Theta}, \mathbf{T}, \mathbf{P}, \mathbf{W}). \qquad (5.13)$$

Basing on these general equations, and taking into account evaporation and evapotraspiration, according to (Choudhury et al., 1994) might be proposed a general equation for description of energy flux from surface site:

$$g_T = f_v g_v + f_s g_s, \qquad (5.14)$$

where f_v and f_s are sites of studied area covered by vegetation or with bare soils. In the most common case, with no ground verification of hydrogeological and geomorphological parameters, according to (Choudhury et al., 1994), might be proposed $f_s = (1 - f_v)$. To more accurate calculation of f_v and f_s, the satellite-based indices may be used (Choudhury et al., 1994):

$$f_v = 1 - \left(\frac{NDVI - NDVI_{\min}}{NDVI_{\max} - NDVI_{\min}} \right)^p, \qquad (5.15)$$

$$f_s = g \cdot \left(\frac{NDWI - NDWI_{\min}}{NDWI_{\max} - NDWI_{\min}} \right)^q. \qquad (5.16)$$

In this algorithm, values of *NDVI* and *NDWI* indices calculating according to (Gao, 1995; Jackson et al., 1983b) in the framework of the observation interval, g, p, q – empirical coefficients.

There are two ways to consider a variability of the energy flux: (i) to construct a special algorithm for spectral indices calculation, which taking into account features of energy balance of the surface, and (ii) application of statistical procedures to ground measurement data analysis aimed to data regularization toward satellite observations. Besides, all data should be temporally regularized.

The task of qualitative and quantitative definition and determination of the spectral intercalibration procedure requires the detailed analysis and processing of huge massive of lab, ground, and remote data. Our overview will be limited by analysis of two spectral indices: *NDVI* (Normalized Difference Vegetation Index) and *NDWI* (Normalized Difference Water Index), which correspond to our task.

Classic equations for calculation of these indices, based on lab experiments, were proposed in (Yu et al., 2010a) for *NDVI* and in (Gao, 1995) for *NDWI*:

$$NDVI^{lab} = \left(\frac{r_{800} - r_{680}}{r_{800} + r_{680}} \right), \tag{5.17}$$

$$NDWI^{lab} = \left(\frac{r_{857} - r_{1241}}{r_{857} + r_{1241}} \right). \tag{5.18}$$

Here r_λ – reflectance in corresponding band λ, nm.

Basing on the balance Equations (5.11–5.14), and taking into account usual bands of satellite sensors, in (Yu et al., 2010b) was proposed algorithms for calculation of these indices using specific sensors:

$$NDVI^{MSS} = \left[\int_{700}^{800} I \, d\lambda - \int_{600}^{700} I \, d\lambda \right] \bigg/ \left[\int_{700}^{800} I \, d\lambda + \int_{600}^{700} I \, d\lambda \right] \bigg/ g. \tag{5.19}$$

For data from MSS sensor of Lansat USGS satellite, and for data from TM and ETM sensors of Lansat satellite:

$$NDVI^{ETM} = \left[\int_{760}^{900} I \, d\lambda - \int_{630}^{690} I \, d\lambda \right]$$
$$\bigg/ \left[\int_{760}^{900} I \, d\lambda + \int_{630}^{690} I \, d\lambda \right] \bigg/ g, \tag{5.20}$$

$$NDWI^{ETM} = \left(\left[\int_{760}^{900} I \, d\lambda - \int_{1550}^{1750} I \, d\lambda \right]$$
$$\bigg/ \left[\int_{760}^{900} I \, d\lambda + \int_{1550}^{1750} I \, d\lambda \right] \right) \bigg/ g. \tag{5.21}$$

Application of reduced-spectral intervals equations allows to obtain distributions of spectral indices reflecting both the specificity of the used sensors and the spatial variations of the energy balance of the earth's surface. Thus, more correct comparison of the results of satellite observations and ground spectrometric measurements may be provided.

Further regularization may be provided by different ways. In the framework of the task solving, the relatively simple way may be proposed. It based on the determination of distribution of studied parameters over the whole area $f_{x,y}$ toward the distribution on studied sites f_m (Fowler et al., 2003; Yu et al., 2015):

$$f_{x,y} = \sum_{m=1}^{n} W_{x,y}(\tilde{f}_m) f_m, \qquad (5.22)$$

where $W_{x,y}(\tilde{f}_m)$ – weighting coefficient, determined as the minimum (Cowpertwait, 1995; Yu et al., 2015):

$$\min \left\{ \sum_{m=1}^{n} \sum_{f_m \in F} w_{x,y}(\tilde{f}_m) \left(1 - \frac{f_m}{\tilde{f}_m} \right)^2 \right\}. \qquad (5.23)$$

In this equation, m – number of measurement points; n – number of measurement series; f_m – distribution of measurement results; F – set of measurement data; $\tilde{f}_m$ – average distribution of measured parameters.

Application of the regularization procedure allows to obtain distributions of the measured parameters, which with controlled reliability correspond to spatial and temporal parameters of satellite data. Comparison of two regular data sets from ground measurements and satellite observations may be done with approach (Acarreta and Stammes, 2005; Cox and Hinkley, 1974), according to:

$$\overline{R} = \int R(\vec{r}) \, d\vec{r}. \qquad (5.24)$$

Here $\vec{r}$ – two-dimensional vector of the site coordinates; R – measured spectral distribution ($\overline{R}$ – with worse spatial resolution).

Thus, ground measurement and satellite observation data can be correctly compared.

5.3 Materials and Data

5.3.1 Satellite Data: Selection, Collection, and Basic Land Cover Classification

According to Equations (5.17) and (5.18) key monitoring parameters are the square and density of the plant cover and the hydrographic network distribution, and key controlled variables are humidity, moisture, photochemical processes intensity, and temperature. So, surface cover classification should be directed to detection of water objects and plant classes. Plant cover, according to Equations (5.16–5.20) might be divided to forests, shrubs, grasslands (natural meadows and farmland), wetlands, and peat bogs with herbaceous vegetation – by the parameters the water balance parameters. Water objects may be divided to lakes, rivers, abandoned channels, river tributaries (natural), water channels (artificial), reservoirs (artificial). Besides, from the viewpoint of risk assessment, i.e., estimation of the possible losses, it is important to identify the elements of the infrastructure: buildings, roads, bridges, dams, etc.

For problem-oriented classification and further analysis of spectral characteristics of terrestrial covers, satellite data USGS Landsat satellites 1975–2009 were used. The list of data used is presented in Table 5.1. As additional materials, the satellite data from EOS Terra MODIS (NOAA), acquired June 3, 2000, August 19, 2004, May 4, 2007, May 22, 2007, May 4, 2009, August 22, 2009, and individual scenes from satellites Ikonos (from 07.10.2006), QuickBird (October 19, 2003), and GeoEye1 (from April 01, 2009 and July 28, 2009) were utilized.

Taking into account the data availability, the basic period for variables analysis was determined a period 1986–2009. The cartographic data of 1972 and 1986 and field verification of CASRE of 2007–2010 were used for data calibration and classification. The hybrid classification procedure with Bayesian maximum likelihood classifier was used (Liu and Mason, 2009).

In Figure 5.1, the result of problem-oriented classification of the area studied is presented.

Legend: 1 – forests (average in period 2003–2009); 2 – shrubs (average in period 2003–2009); 3 – natural meadows, pastures (average in period March–July 1999–2009); 4 – agricultural lands (average in period May–July 2007–2009); 5 – wetlands, overmoistured surface sites (average in period March–July 1999–2009); 6 – peatland (average in period 1999–2009); 7 – natural water objects (on May 2009); 8 – artificial water objects (on May 2009); 9 – built areas (on May 2009); 10 – infrastructure: roads, bridges (in 2003, 2007, and 2009).

Further analysis of the data was directed to study of individual spectral characteristics of certain types of terrestrial cover, taking into account the existing tendencies of changes in local (in particular climatic) parameters.

Table 5.1   List of satellite data used to analysis of the study area

Satellite	Sensor	Date	Purpose
Landsat-2	MSS	June 11, 1975	Land-cover classification, spectral indices control (*NDVI*)
Landsat-2	MSS	March 19, 1976	Land-cover classification, spectral indices control (*NDVI*)
Landsat-2	MSS	May 12, 1976	Land-cover classification, spectral indices control (*NDVI*)
Landsat-2	MSS	June 24, 1976	Land-cover classification, spectral indices control (*NDVI*)
Landsat-5	TM	May 04, 1986	Land-cover classification, spectral indices control (*EVI, NDVI*)
Landsat-5	TM	July 07, 1986	Land-cover classification, spectral indices control (*EVI, NDVI, NDWI, SIPI, PRI, PSI*)
Landsat-7	ETM	July 19, 1999	Land-cover classification, spectral indices control (*EVI, NDVI*)
Landsat-7	ETM	February 28, 2000	Land-cover classification, spectral indices control (*EVI, NDVI*)
Landsat-7	ETM	May 02, 2000	Land-cover classification, spectral indices control (*EVI, NDVI*)
Landsat-7	ETM	March 24, 2003	Land-cover classification, spectral indices control (*EVI, NDVI*)
Landsat-7	ETM	May 27, 2003	Land-cover classification, spectral indices control (*EVI, NDVI*)
Landsat-5	TM	July 22, 2003	Land-cover classification, spectral indices control (*EVI, NDVI, NDWI, SIPI, PRI, PSI*)
Landsat-5	TM	April 28, 2007	Land-cover classification, spectral indices control (*EVI, NDVI*)
Landsat-5	TM	May 14, 2007	Land-cover classification, spectral indices control (*EVI, NDVI*)
Landsat-7	ETM	May 22, 2007	Land-cover classification, spectral indices control (*EVI, NDVI*)
Landsat-5	TM	June 15, 2007	Land-cover classification, spectral indices control (*EVI, NDVI*)
Landsat-7	ETM	July 09, 2007	Land-cover classification, spectral indices control (*EVI, NDVI, NDWI, SIPI, PRI, PSI*)
Landsat-5	TM	July 17, 2007	Land-cover classification, spectral indices control (*EVI, NDVI, NDWI, SIPI, PRI, PSI*)
Landsat-7	ETM	April 09, 2009	Land-cover classification, spectral indices control (*EVI, NDVI*)
Landsat-7	ETM	April 25, 2009	Land-cover classification, spectral indices control (*EVI, NDVI*)

(*Continued*)

Table 5.1   (Continued)

Satellite	Sensor	Date	Purpose
Landsat-7	ETM	May 27, 2009.	Land-cover classification, spectral indices control (*EVI, NDVI*)
Landsat-7	ETM	June 28, 2009	Land-cover classification, spectral indices control (*EVI, NDVI*)
Landsat-7	ETM	July 14, 2009	Land-cover classification, spectral indices control (*EVI, NDVI, SIPI, PRI, PSI*)

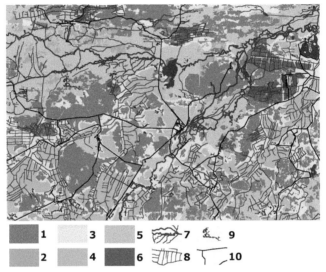

Figure 5.1   Result of the problem-oriented classification of the area studied: basins of the Stokhid and Prypyat rivers, north-western part of Ukraine.

5.3.2 Satellite Data Analysis: Spectral Processing

Proposed risk assessment method based on estimation of ecosystems stresses generated by variations of water and heat dynamics. Analysis of ecosystems reaction is based on detection of spectral response on the set of spectral indices (Blackburn, 1998; Choudhury, 2001; Dobrowski et al., 2005; Penuelas et al., 1995 and Verma et al., 1993).

The set of spectral indices has been calculated to estimation of stresses using the Landsat data (see Table 5.1) for the study region (upper sites of Western Bug river, areas 20 × 20 km with coordinates of the center: 490 57 '15,36" N, 240 46 '05,31" E). Data were compared for the entire available period separately for forest (crown) and grass vegetation (natural meadows

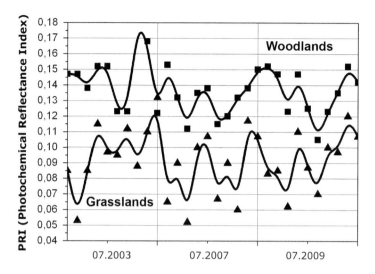

Figure 5.2 *PRI* index distribution.

and wetlands), according to the previous classification of satellite images (see Figure 5.1).

The determined tendencies of the spectral indices dynamics indicate the presence of significant water and less exhibited temperature stress. The analysis of long-term water, temperature, and radiation stresses that would result a noticeable reaction of regional ecosystems can be accomplished by analyzing the changes in vegetation photosynthetic activity, estimated by *PRI* (Photochemical Reflectance Index) for the studied region (Figure 5.2).

Some decreasing of *PRI* values over study area is observed for the forested sites from 0.145 to 0.13 (about 10% to average). But for grass sites *PRI* demonstrates some increasing: from 0.09 to 0.095 (about 5.4% to average). Taking into account significant variations of PRI during the observation period (about 25% for grasslands and about 14% for forested areas), available observations not allow to detect any stable trends of the studied index.

PRI index was calculated for TM sensor according to algorithm (Gamon et al., 1997) in the form (Yu et al., 2010a):

$$PRI^{TM} = \left[\int_{0.45}^{0.52} I\, d\lambda - \int_{0.52}^{0.60} I\, d\lambda \right] \bigg/ \left[\int_{0.45}^{0.52} I\, d\lambda + \int_{0.52}^{0.60} I\, d\lambda \right] \bigg/ g \,,$$

$$(5.25)$$

And for *ETM* sensor as (Yu et al., 2010a):

$$PRI^{TM} = \left[\int_{0.45}^{0.52} I \, d\lambda - \int_{0.52}^{0.60} I \, d\lambda\right] \Big/ \left[\int_{0.45}^{0.52} I \, d\lambda + \int_{0.52}^{0.60} I \, d\lambda\right] \Big/ g \,.$$

(5.26)

Here λ – is the wavelength, g – semiempiric gain factor, in this case $g = 600$.

PRI data show that during the observation period in the study region did not significant impact, which may lead to change of species composition or vegetation cycles.

For detection of stress-related changes of the vegetation the distributions of *NDVI* (Normalized Difference Vegetation Index) (Jackson et al., 1983b) and *EVI* (Enhanced Vegetation Index) indices has been analyzed (Huete et al., 1997). These indices may be used as the indicators of impact of long-term landscape changes. To detect an impact of water stresses a *SIPI* (Structure Intensive Pigment Index) index was analyzed (Yu et al., 2012). According to (Yu et al., 2010) these indices were calculated as follow:

$$NDVI^{MSS} = \left[\int_{0.70}^{0.80} I \, d\lambda - \int_{0.60}^{0.70} I \, d\lambda\right] \Big/ \left[\int_{0.70}^{0.80} I \, d\lambda + \int_{0.60}^{0.70} I \, d\lambda\right] \Big/ g \,,$$

(5.27)

For MSS sensor of Landsat 1–5 satellites; and for TM and ETM sensors of Landsat 4–7 satellites:

$$NDVI^{TM/ETM} = \left[\int_{0.76}^{0.90} I \, d\lambda - \int_{0.63}^{0.69} I \, d\lambda\right]$$
$$\Big/ \left[\int_{0.76}^{0.90} I \, d\lambda + \int_{0.63}^{0.69} I \, d\lambda\right] \Big/ g \,.$$

(5.28)

Where a gain factor $g = 200$.

$$EVI^{TM} = 3.2 \left[\int_{0.76}^{0.90} I \, d\lambda - \int_{0.63}^{0.69} I \, d\lambda\right] \Big/ \left[\int_{0.76}^{0.90} I \, d\lambda\right.$$
$$\left. +6\int_{0.63}^{0.69} I \, d\lambda - 7.5\int_{0.45}^{0.52} I \, d\lambda + 1\right] \Big/ g,$$

(5.29)

$$EVI^{ETM} = 2.5 \left[\int_{0.760}^{0.900} I \, d\lambda - \int_{0.630}^{0.690} I \, d\lambda\right] \Big/ \left[\int_{0.760}^{0.900} I \, d\lambda\right.$$
$$\left. + 6\int_{0.630}^{0.690} I \, d\lambda - 7.5\int_{0.450}^{0.515} I \, d\lambda + 1\right] \Big/ g.$$

(5.30)

Here gain factor $g = 500$. Besides, some simplified algorithm can be proposed for the *EVI* index (Huete et al., 1997):

$$EVI^{TM/ETM} = 2.5 \left[\int_{0.76}^{0.90} I\, d\lambda - \int_{0.63}^{0.69} I\, d\lambda \right]$$
$$\left/ \left[\int_{0.76}^{0.90} I\, d\lambda + 2.4 \int_{0.63}^{0.69} I\, d\lambda + 1 \right] \right/ g , \quad (5.31)$$

Here $g = 500$.

$$SIPI^{TM} = \left[\left[\int_{0.76}^{0.90} I\, d\lambda - \int_{0.45}^{0.52} I\, d\lambda \right] \right.$$
$$\left/ \left[\int_{0.76}^{0.90} I\, d\lambda - \int_{0.63}^{0.69} I\, d\lambda/g \right] - 1 \right. , \quad (5.32)$$

$$SIPI^{ETM} = \left[\left[\int_{0.760}^{0.900} I\, d\lambda - \int_{0.450}^{0.515} I\, d\lambda \right] \right.$$
$$\left/ \left[\int_{0.760}^{0.900} I\, d\lambda - \int_{0.630}^{0.690} I\, d\lambda/g \right] - 1 \right. . \quad (5.33)$$

Here $g = 50$ (*SIPI* = 0, if *SIPI* < 0).

Direct water stress was estimated by *NDWI* (Normalized Difference Water Index) index (Gao, 1995), according to algorithm (Yu et al., 2010):

$$NDWI^{TM/ETM} = \left(\left[\int_{0.76}^{0.90} I\, d\lambda - \int_{1.55}^{1.75} I\, d\lambda \right] \right.$$
$$\left/ \left[\int_{0.76}^{0.90} I\, d\lambda - \int_{1.55}^{1.75} I\, d\lambda/g \right] \right) \left/ g. \right. \quad (5.34)$$

Where gain factor $g = 100$. Proposed forms (5.25–5.34) may be applied also for field measurements in corresponding spectral bands.

Figures 5.3–5.5 demonstrate distributions of calculated indices.

For the period 1975–2009 in the study area it may be observed increasing of *NDVI* values from 0.45 to 0.52 (about 15% of average value 0.47) for the grasslands, including pastures and wetlands; and for forested areas this index increased from 0.5 to 0.651 (about 27% of average 0.552). Variations of index are 12% for grasslands and 9% for woodlands.

For the period 1986–2009, index *EVI* demonstrates increasing from 0.175 to 0.218 (about 21.8% to average 0.043), and decreasing for grasslands

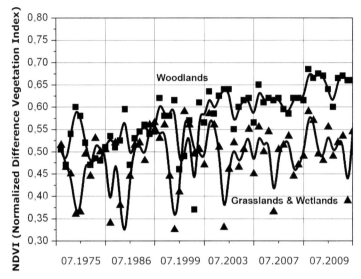

Figure 5.3   Distribution of *NDVI* index for the study area.

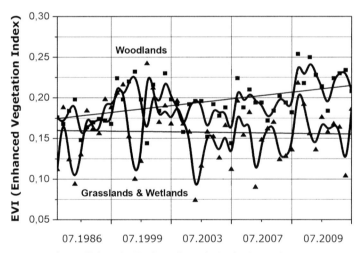

Figure 5.4   Distribution of *EVI* index in the study area.

from 0.16 to 0.155 (3.2% to average 0.155). Index variations are 3% for grasslands and for woodlands. So, *EVI* increased in woodlands and is constant in grasslands.

Pigmentation is an important indicator of plant stress. So it is interesting to analyze a *SIPI* index (Figure 5.5).

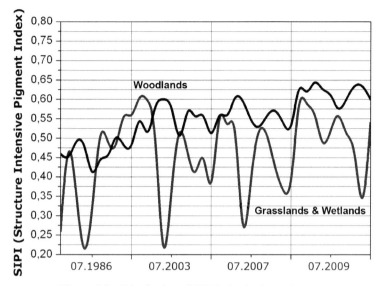

Figure 5.5 Distribution of *SIPI* index in the study area.

In the period 1986–2009, index *SIPI* demonstrates increasing from 0.42 to 0.5 (19% to average 0.43) for grasslands, and from 0.46 to 0.63 (30% to average 0.56) increasing for forested areas.

So it is possible to conclude existence of clearly manifested water stress, which corresponds to increasing of precipitation increasing (30%) in this period in the study area. Less values of stress in the grasslands connected with artificial drainage of these sites.

The determined changes of the spectral indices can be used as the parameters in method of assessment of risks of the processes that cause stress associated with changes of spectral characteristics of the vegetation.

5.3.3 Satellite Data Calibration using the In-Field Spectrometry Measurements

Field spectrometric measurements have been carried out aimed to calibration of described spectral indices. In the separated field sites of study area using spectrometer FieldSpec®3 FR in period 2007–2015 were acquired 117 spectral signatures in different ranges: in range 350–2,500 nm with bandwidth 1.4 nm, in the range 350–1,000 nm with bandwidth 2 nm, in range 1,000–2,500 with bandwidth 3 nm, and in range 1,400–2,100 nm with bandwidth 10 nm.

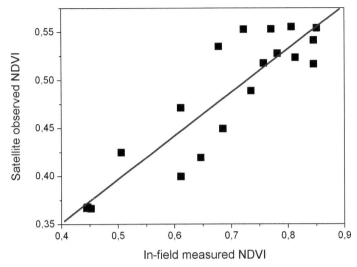

Figure 5.6   Result of comparison of *NDVI* indices from satellite and field data, calculated by "ETM" algorithm ($R = 0.90$; $\sigma = 0.031$).

Because direct comparison of field and satellite data is incorrect through different methodology, energetic, spatial, and temporal characteristics of acquiring, the regularization procedures have been applied.

Regularization by energy flux (spectral regularization) (5.18–5.21) is necessary to obtain data distributions with stable intercorrelation.

The best correlations may be found between data, calculated using "ETM" algorithm (5.20) and (5.21) (see Figure 5.6).

The better correlations between satellite and field data may be obtained using the spatial-temporal regularization algorithms (5.22–5.24).

To calibrate a data of satellite observations the calibrations correlations in the form of linear regressions may be proposed.

$$NDVI_{sat} = 0.22 + 0.38 \cdot NDVI_{ground}^{lab}, \qquad (5.35)$$

$$NDVI_{sat} = 0.19 + 0.55 \cdot NDVI_{ground}^{MSS}, \qquad (5.36)$$

$$NDVI_{sat} = 0.17 + 0.45 \cdot NDVI_{ground}^{ETM}. \qquad (5.37)$$

These formulas can be applied as the calibration dependencies for vegetation spectral indices.

The same way may be proposed for the water indices *NDWI*.

It should be noted that *NDWI* indices has better correlations than vegetation indices, although its spatial resolution is worse (see Figure 5.7).

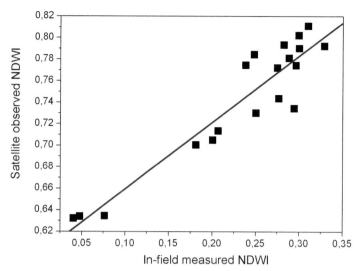

Figure 5.7 Result of comparison of *NDWI* indices from satellite and field data, calculated by "ETM" algorithm ($R = 0.94$; $\sigma = 0.021$).

Linear approximation equations for average calculated satellite *NDWI* index and field measured data also may be presented:

$$NDWI_{sat} = 0.65 + 2.75 \cdot NDWI_{ground}^{lab}, \qquad (5.38)$$

$$NDWI_{sat} = 0.58 + 0.61 \cdot NDWI_{ground}^{ETM}, \qquad (5.39)$$

$$NDWI_{sat} = 0.29 + 0.61 \cdot NDWI_{ground}^{ETM}. \qquad (5.40)$$

So, the procedure of calibration of satellite derived spectral indices on the base of field spectrometry is constructed.

It need to be noted that there is a substantial difference between satellite and field spectrometry, connected with scheme of survey. Field spectrometry uses limbic scheme of observation, which can leads to losses of information under anisotropic landscapes. Satellite surveys use nadir scheme, which collect full signal from the site. So, signals collected can be different. To reduce this difference, it is necessary to calculate general function of view of field spectrometer. It may reduce corresponding uncertainty to 12%–15%. Generally speaking, it is necessary to determine in the explicit form the distribution of angles field spectrometric survey $p_m(\vartheta_i)$, where ϑ – spectrometer view angle, m – measurements sites, i – measurements points, as a function $p_m(\vartheta_i) \rightarrow \mathbf{F}(\theta(t), x_i, y_i, z_i, z_i^*)$, where θ – solar angle, t – measurement time, x, y – coordinates of measurement point, z – terrain, $z*$ – "effective height of plants" (position of the spectrometer relative to the vegetation surface). It is necessary to reduce the uncertainties and errors, connected with differences of limbic and nadir survey schemes.

Besides, it should be specially noted that determined linear intercalibration dependencies for *NDVI* index (5.35–5.37) are enough correct only in the interval of *NDVI* values 0.4–0.55. If *NDVI* > 0.55 calibration dependency will not be linear, as it show the analyzed data. Determination of intercalibration dependencies for whole interval of possible values of the index is the task of future studies.

Therefore, it was obtained important tool for quantitative calibration not only satellite observation data, but also its algebraic compositions, such as the spectral indices, which are necessary for assessment of parameters of hydrological risks.

5.3.4 Meteorological and Climatic Data Analysis

In the framework of the solving of the task of the assessment of hydrological risks the climate impact to hydrological parameters should be estimated. In particular, local distributions of climatic and meteorological parameters in the observation period over the study area should be analyzed. It is necessary to construct the statistical significant distributions of parameters, which are the variables of basic model Equations (5.3–5.7) and (5.10).

Data of long-term meteorological observations of local stations were used for this study GSOD (2019) (see Table 5.2).

Totally 137,948 meteorological records was analyzed, 75,717 from which correspond to the observation period. Trends of maximal, mean, and minimal values of studied parameters, correspond to vegetation cycle, were detected (see Figures 5.8 and 5.9).

These data show that average monthly temperature T_{mean} in the study region in July decreased to 1.55°C – from 21.75°C to 20.2°C. It reflects to productivity of landscapes, which not demonstrate any significant increasing of biomass, although annual mean temperature T_{mean}^{annual} increase to 1.25°C – from 7.25°C to 8.5°C.

Extreme – maximal and minimal – temperatures also increased. Minimal recorded temperature T_{min} increased to 1.75°C from 6.5°C to 8.25°C, and maximal T_{max} increased to 1.42°C – from 30.78°C to 32.2°C. These drastic changes of extreme temperature lead to increasing of extreme events frequency and intensity more than changes of mean temperature.

Table 5.2   List of meteorological stations used for study of local distributions of climatic parameters

Location, town	Name (WMO)	Descriptor (USAF)	lat	long	alt, m
				Coordinates	
Volodymyr-Volynsky (Ukraine)	Volodymyr-Volynskyi	331770	50.833	24.317	194
Lviv (Ukraine)	Lviv	333930	49.817	23.950	323
Rivne (Ukraine)	Rivne	333010	50.583	26.133	231
Sarny (Ukraine)	Sarny	330880	51.283	26.617	156
Siedlce (Poland)	Siedlce	123850	52.250	22.250	155
Lublin (Poland)	Lublin-Radawiec	124950	51.217	22.400	240
Brest (Byelorussia)	Brest	330080	52.117	23.683	143
Pinsk (Byelorussia)	Pinsk	330190	52.117	26.177	142

At the same time, during the observation period the changes of precipitation were significant: during 1985–2010 increase of precipitation is about 25%.

Detected changes of climatic parameters allow to conclude that in study region water stresses, connected with change of precipitation, are more significant than heat stresses, generated by temperature changes.

5.4 Method of the Risk Assessments using Bayes Approach

Because the input satellite data have a stochastic nature, a probability of stress detection using the set of spectral indicators can be assessed by Bayes rule (Bernardo et al., 1994) as:

$$P(\Delta SRI^*(x,y)|Q_{stress}) = \frac{P_s(x,y) \cdot \prod_N P_N(\Delta SRI^*|Q_{stress})}{\int_{x,y} P_N(\Delta SRI^*|Q)dP_s(x,y)}$$

$$= \frac{P_S(x,y) \cdot P_N(\Delta SRI^*|Q_{stress})}{P_N(\Delta SRI^*|Q_{stress})P_S(x,y) + P_N(\Delta SRI^*|Q_0)P_0(x,y)}. \quad (5.41)$$

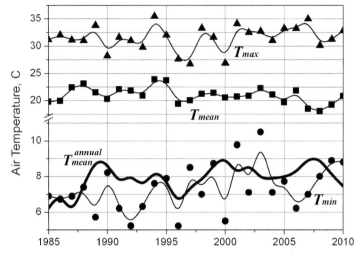

Figure 5.8   Monthly average (July) T_{mean}, annual mean T_{mean}^{annual}, maximal T_{max}, and minimal T_{min} recorded air temperatures in the study region in the period 1985–2010.

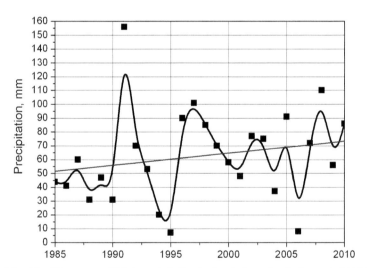

Figure 5.9   Monthly average (July) precipitation in the study region in the period 1985–2010.

Here $\Delta SRI*$ is the period-reduced spectral index (normalized difference between observation period average and detected value at observation moment), which used for analysis (*NDVI, EVI, NDWI, SIPI*), index Q_{stress} marks sites impacted by stresses, index Q_0 marks classes of pixels, where stresses are not present with high reliability. Probability $P_S(x, y)$ is determining self-empirically. Ratio of probabilities $P_S(x, y)$ and $P_0(x, y)$ is determining as: $\lim_{x,y,\tau}(P_S(x, y)_\tau + P_0(x, y)_\tau) = 1$ (i.e., on the enough long periods it is possible to assume that: $P_S(x, y) = 1 - P_0(x, y)$). Probability $P_S(x, y)$ can be determined using Gauss weight function, if pixel on the satellite scene does not refer to the site where stress factors are observed and cannot be unambiguously attributed to site, in which there is no stress. So, the probability of uncertain presence of stress at location (that corresponds to position of this pixel on the surface) will be a function of the geometric distance from the closest place under the registered stress:

$$P_S(x, y) = P_{\min} + (P_{\max} - P_{\min}) \cdot e^{d_s^2 / 2\sigma_p^2}. \tag{5.42}$$

Where $P_S(x, y)$ – probability of existence (or appearance in time scale of observation period) of the studied stresses; P_{max} – the maximum probability of the current presence of stress in the study site (unregistered during the interpretation and classification of the image), which depends on the type of sensor, the physical and geographical features of the region and the type of surface (based on the general methodological principles (Ermoliev and Winterfeldt, 2010), P_{max} for TM and ETM sensors of Landsat satellite may be assessed as $0.25 - 0.28$); P_{min} – minimum probability, depending on sensor type, physical and geographical features of the region, and type of surface (P_{min} may be assessed about 0.01); $d_s(x,y)$ – distance from the nearest place under the registered stress; σ_p – an empirical parameter to be determined, using field research data, basing the characteristics of vegetation cover of the study area and the type of sensor (for example, for TM and ETM sensors of Landsat satellites in the study region, the parameter σ_p can be assessed as 1.1–1.5 km). Therefore, for the study region, for TM and ETM sensors of Landsat satellite the parameter $P_S(x, y)$ can be calculated according to simplified algorithm:

$$P_S(x, y) = 0.01 + 0.26 \cdot e^{d_s^2 / 1.69}. \tag{5.43}$$

Calculated probabilities may be used as the basis for assessment of risks, connected with hydrological and hydrogeological threats. Assessment of complex risk is a complicated task and requires taking into account multiscale processes and links in studied multicomponent systems (Ermoliev, 2009). In the general case, the following equation may be proposed (Jones et al., 1996):

$$R(t) = f_A(R^L(t), R_0(t)) \iint_{xy} \int_{t_0}^{t} p(v) f(x, y, v) \, dt \, dx \, dy. \tag{5.44}$$

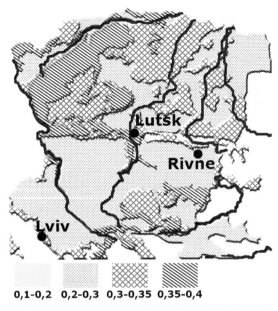

Figure 5.10   The risk of under-flooding calculated for the study region.

Here $f_A(R^L(t), R_0(t))$ is the approximation function of impact, which describes interaction of short- and long-term impact factors to complex hydrological and hydrogeological security (depending of the used model, it may be presented in different form, even as linear composition of corresponding probabilities; $p(v)$ – probability of negative impact with determined conditions, where v – effective velocity of development of the disastrous process (modeled by flood Equations (5.2) and (5.2) and the model of run-off); $R_0(t)$ – general (mean) probability of disaster (in general case this is the probability distribution function of impacts $f^\alpha(\psi, I)$ over the site $\psi(x, y)$; function of losses – parametric description of negative impact $f_v(I)$; and risk function H, which determined from regional disasters statistics (Yu et al., 2015)), and might be calculated using remote sensing data; $R^L(t)$ – long-term risk (here should be used the impact factors Q_j from the set $j \in \Im$, connected with climate and environmental changes, and scenarios of change of disasters frequency $\mathbf{F}(Q_j)$ calculating with forecast models (Climate Change, 2001; Anthes et al., 1987; Yu , 2014)); $f(x,y,v)$ – function of expansion, describing by distribution of hydrological (properties of water-table) and geomorphological (location and size of water discharge zones) parameters using (5.4–5.9).

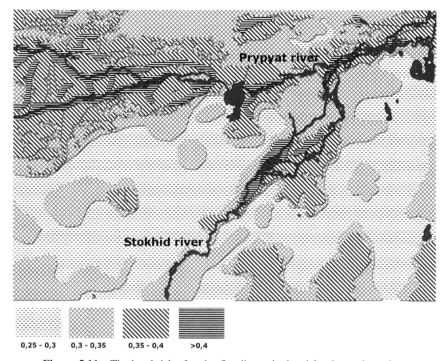

Figure 5.11 The local risk of under-flooding calculated for the study region.

It should be noted that expansion function $f(x,y,v)$ may also be determined through satellite data classification and its verification with field data (Earth, 2012).

Using the approach proposed (5.44) and the data of satellite and field research described, the hydrological risks in study region can be assessed. The result of this assessment is presented in Figures 5.10 and 5.11.

Using this approach the current state of hydrological security and the current level corresponding risks can be assessed for the certain period of time. But the analysis of the long-term behavior of the investigated variables the study of a wider set of parameters – both climatic and environmental, which should be used in predictive models – is required.

5.5 Conclusions

The presented results allow to conclude that based on satellite observation data. Bayes risk assessment techniques are an adequate basis for assessing the hydrological and hydrogeological risks both at the regional and local levels. Based on the model of hazardous processes, it is possible to construct a method for processing of the satellite images and analyzing the spectral indicators.

Basing on the data of 1975–2009, the sets of the spectral indices (*EVI, NDVI, NDWI, SIPI, PRI*) that could be used as stress indicators of the ecosystems studied and connected with the corresponding risks were identified. The form of algorithm for calculation of indices both for satellite data and for field spectrometric measurements obtained in 2007–2015 is proposed. The results of calculations of spectral indices were compared with the regional distribution of climatic parameters over the observation period; key factors that influencing to the hydrological and hydrological security, as well as the impacts on ecosystems associated with changes in climatic indicators were determined.

To determine probability of stress detection by the set of spectral indices, the appropriate algorithm was proposed, values of key variables for specified parameters of sensors and regional landscape conditions were determined. The method of estimation of complex risk connected with hydrologic and hydrogeological threats was offered. The risks of flooding on the investigated territory are calculated and mapped.

It should be noted that quantitative risk assessment (which is finally is an economic category) in the long-term perspective requires a certain strategy of socioeconomic development, presented in terms of quantitative indicators, which currently does not exist at present, therefore the reliability of forecast and scenarios is significantly reduced.

Particular attention to the phenomenon of swamping is required. Swamps in the studied area are the integral part of regional ecosystems. Wetlands are widely represented in the region. The processes associated with the swamps, including the economically negative, are controlling with the drainage network quite successfully. Significant changes of the swamps square during the observation period in the studied region were not detected. However, changes in the area of crown vegetation (in particular, the registered declining forest area (Movchan et al., 2014) 45) and nonoptimal agricultural management can have a negative impact on the water balance and increase the risk of swamping. So, according to the results obtained, the risk of swamping can be assessed as not very high. But, taking into account climatic changes, including its regional and local scale, it is necessary to take into account ecological changes in wetlands, such as risks of intrusive species development.

In practice, the results obtained may be used for the assessing and forecasting the risks of flooding and swamping of the ecologically sensitive areas, for development of decision-making systems for losses minimization, as well as for development of ecological monitoring systems.

Acknowledgments

The authors are grateful to anonymous referees for constructive suggestions that resulted in important improvements to the chapter, to colleagues from the International Institute for Applied Systems Analysis (IIASA), from the American Statistical Association (ASA), American Meteorological Society (AMS), and from the International Association for Promoting Geoethics (IAPG) for their critical and constructive comments and suggestions. Particular thanks authors express to Science and Technology Center in Ukraine (STCU) for partial support of this study in the framework of research project #6165 "Information and Technological Support for Greenhouse Effect Impact Assessment on Regional Climate using Remote Sensing."

References

Acarreta, J. R. and Stammes, P. (2005). 'Calibration comparison between SCIAMACHY and MERIS onboard ENVISAT', IEEE Geosci. Remote Sens. Lett., 2(1), 31–35.

Anthes, R. A., Hsie, E. Y. and Kuo, Y. H. (1987). 'Description of the Penn State/NCAR Mesoscale Model Version 4(MM4)//NCAR', Techn. Note, NCAR/TN-282STR, Natl Cent. For Atmos. Res., Boulder, Colo, 70.

Bartell, S. M., Gardner, R. H., and O'Neill, R. V. (1992). 'Ecological risk estimation'. Boca Raton, StateFL: Lewis Publishers.

Bernardo, J. M. and Smith. A. F. M. (1994). 'Bayesian Theory'. Chichester, UK: Wiley.

Blackburn, G. A. (1998). 'Spectral indices for estimation photosynthetic pigment concentrations: A test using senescent tree leaves', Int. J. Remote Sensing, 4, 657–675.

Budyko, M. I. (1974). 'Climate and Life', New York: Elsevier, 508.

Castelli, F., Entekhabi, D. and Caporali, E. (1999). 'Estimation of surface heat flux and an index of soil moisture using adjoint state surface energy balance', Water Resour. Res., 35, 3115–3125.

Choudhury, B. J., Ahmed, N. U., Idso, S. B., Reginato, R. J. and Daughtry, C. (1994). 'Relations between evaporation coefficients and vegetation indices studied by model simulations', Remote Sens. Environ., 50, 1–17.

Choudhury, B. J. (2001). 'Estimating gross photosynthesis using satellite and ancillary data: Approach and preliminary results'. Remote Sensing of Environment, 75, 1–21.

'Climate Change 2001. The Scientific Basis. Report of the Inergovernmental Panel on Climate Change', UNEP, WMO, 2001.

Cowpertwait, P. S. P. (1995). 'A generalized spatial-temporal model of rainfall based on a clustered point process', Proc. R. Soc. Lond. Ser. A, 450, 163–175.

Cox, D. R. and Hinkley, D. V. (1974) 'Theoretical Statistics', New York: Chapman & Hall, 285.

Dobrowski, S. Z., Pushnic, J. C., Zarco-Tejada, P. J. and Ustin, S. L. (2005). 'Simple reflectance indices track heat and water stress-induced changes in steady-state chlorophyll fluorescence at the canopy scale', Remote Sensing of Environment, 97, 403–414, doi:10.1016/j.rse.2005.05.006.

'Earth Systems Change over Eastern Europe', (2012). Lyalko, V. and Groisman, P. (eds.), Kiev: Academperiodyka Press.

Entekhabi, D., Rodriguez-Iturbe, I. and Bras, R. (1992). 'Variability in large scale water balance with land surface-atmosphere interaction', J. Clim., 5, 798–813.

Ermoliev, Y. and Winterfeldt von, D. (2010). 'Risk, security and robust solutions', IIASA Interim Report, IR-10-013, IIASA, 41.

Ermoliev, Y. (2009). 'Stochastic quasigradient methods: Applications', In C. Floudas and P. Pardalos, (eds.), Encyclopedia of Optimization. New York: Springer Verlag, 3801–3807.

Ermoliev, Y. and Hordijk, L. (2006). 'Global changes: Facets of robust decisions', In. 'Coping with Uncertainty, Modeling and Policy Issues'. K. Marti, Y. Ermoliev, M. Makowski, G. Pflug, (eds.), Berlin, Germany: Springer-Verlag, 4–28.

Fenton, N. and Neil, M. (2012). 'Risk assessment and decision analysis with Bayesian networks', CRC Press, Boca Raton, FL.

Fisher, S. G. and Woodmansee, R. (1994) 'Issue paper on ecological recovery', in 'Ecological risk assessment issue papers', Washington, DC: Risk Assessment Forum, U. S. Environmental Protection Agency, 7-1–7-54. EPA/630/R-94/009.

Fowler, H. J., Kilsby, C. G. and O'Connell, P. E. (2003). 'Modeling the impacts of climatic change and variability on the reliability, resilience and vulnerability of a water resource system', Water Resour. Res., 39, 1222, 2003.

Gamon, J. A., Serrano, L. and Surfus, J. S. (1997). 'The Photochemical Reflectance Index: An optical indicator of photosynthetic radiation use efficiency across species, functional types and nutrient levels', Oecologia, 112, 492–501.

Gao, B. C. (1995). 'Normalized difference water index for remote sensing of vegetation liquid water from space', Proceedings of SPIE, 2480, 225–236.

'GSOD (Global Summary of the Day)'. National Climatic Data Center of U.S. Department of Commerce, courtesy of NOAA Satellite and Information Service of National Environmental Satellite, Data, and Information Service (NESDIS) online at www.ncdc.noaa.gov.

Guha-Sapir, D., Vos, F., Below, R. and Ponserre, S. (2011). 'Annual Disaster Statistical Review 2010: The numbers and trends', Brussels, Universit catholique de Louvain, Centre for Research on the Epidemiology of Disasters (CRED) 50.

Gupta, H. V., Bastidas, L. A., Sorooshian, S., Schuttleworth, W. J. and Yang, Z. L. (1999). 'Parameter estimation of a land surface scheme using multicriteria methods', J. Geophys. Res., 104, 19491–19503.

Huete, A. R., Liu, H., Batchily, K. and van Leeuwen, W. (1997). 'A comparison of vegetation indices over a global set of TM images for EOS-MODIS', Remote Sensing of Environment, 59(3), 440–451.

Liu, J. G. and Mason, Ph. J. (2009). 'Essential image processing and GIS for remote sensing'. CityOxford, Imperial College London, UK: John Wiley & Sons, 462, ISBN: 978-0-470-51032-2.

Jackson, R. D., Slater, P. N. and Pinter, P. J. (1983). 'Discrimination of growth and water stress in wheat by various vegetation indices through clear and turbid atmospheres', Remote Sens. Environ., 15, 187–208.

Jackson, R. D., Slater, P. N. and Pinter, P. J. (1983). 'Discrimination of growth and water stress in wheat by various vegetation indices through clear and turbid atmospheres', Remote Sens. Environ., 15, 187–208.

Jiang, X. and Mahadevan, S. (2007). 'Bayesian risk-based decision method for model validation under uncertainty', Rel. Eng & Sys. Safety, 92(6), 707–718.

Jones, R. G., Murphy, J. M., and Noguer, M. (1996). 'Simulation of climate change over Europe using nested regional climate model. I: Assessment of control climate, including sensitivity to location of lateral boundaries'. Quart. J Roy. Meteorol. Soc., 77, 1413–1449.

Krayenhoff van de Leur, D. A. (1958). 'A study of non-steady groundwater flow with special reference to a reservoir coefficient', Ingenieur, 70, 87–94.

Movchan, D., Kostyuchenko, Yu. V., Marton, L., Frayer, O. and Kyryzyuk, S. (2014). 'Uncertainty Analysis in Crop Productivity and Remote Estimation for Agricultural Risk Assessment', in Vulnerability, Uncertainty, and Risk: Quantification, Mitigation, and Management, by Michael Beer, Siu-Kui Au, Jim W. Hall, (eds.), 1008–1015, Liverpool, UK: ASCE, 2014. doi: 10.1061/9780784413609.102.

Penuelas, J., Baret, F. and Filella, I. (1995). 'Semi-empirical indices to assess carotenoids/chlorophyll a ratio from leaf spectral reactance'. Photosynthetica, 31, 221–230.

Peters-Lidard, C. D., Zion, M. S. and Wood, E. F. (1997). 'A soil – vegetation – atmosphere transfer scheme for modeling spatially variable water and energy balance processes', J. Geophys. Res., 102, 4303–4324.

Regulation (EU) No. 911/2010 of the European Parliament and the Council on the European Earth monitoring programme (GMES) and its initial operations (2011–2013) // Official Journal of the European Union, 20.10.2010, L 276/1–L 276/10C.

Rodriguez-Iturbe, I., Entekhabi, D. and Bras, R. L. (1991). 'Nonlinear dynamics of soil moisture at climate scales: 1. Stochastic analysis', Water Resour. Res., 27(8), 1899–1906.

'The Use of Earth Observing Satellites for Hazard Support: Assessments & Scenarios', CEOS/NOAA, 2001, 218.

Verma, S. B., Sellers, P. J., Walthall, C. L., Hall, F. G., Kim, J. and Goetz, S. J. (1993). 'Photosynthesis and stomatal conductance related to reflectance on the canopy scale'. Remote Sensing of Environment, 44, 103–116, 1993.

Yu. Kostyuchenko, V. (2014). 'Infrastructure vulnerability assessment toward extreme meteorological events using satellite data', in 'Numerical Methods for Reliability and Safety Assessment: Multiscale and Multiphysics Systems', S. Kadry and A. El Hami, (eds.), Switzerland: Springer International Publishing.

Yu. Kostyuchenko, V., Movchan, D., Artemenko, I. and Kopachevsky, I. (2017). 'Stochastic Approach to Uncertainty Control in Multiphysics Systems: Modeling of Carbon Balance and Analysis of GHG Emissions Using Satellite Tools', in 'Mathematical Concepts and Applications in Mechanical Engineering and Mechatronics', M. Ram and J. P. Davim, Eds., USA: IGI Global, 350–378.

Yu. Kostyuchenko, V., Kopachevsky, I., Yuschenko, M., Solovyov, D., Marton, L. and Levynsky, S. (2012). 'Spectral reflectance indices as indirect indicators of ecological threats', in 'Sustainable Civil Infrastructures – Hazards, Risk, Uncertainty', K. K. Phoon, M. Beer, S. T. Quek, S. D. Pang, (eds.), Singapore: Research Publishing, 557–562, 2012.

Yu. Kostyuchenko, V., Kopachevsky, I., Solovyov, D., Bilous, Y. and Gunchenko, V. (2010a). 'Way to reduce the uncertainties on ecological consequences assessment of technological disasters using satellite observations', Proc. of the 4th International Workshop on Reliable Engineering Computing "Robust Design – Coping with Hazards, Risk and Uncertainty", March 3–5, Singapore, National University of Singapore, 765–776.

Yu. Kostyuchenko, V., Kopachevsky, I., Solovyov, D., Bilous, Y. and Gunchenko, V. (2010b). 'Way to reduce the uncertainties on ecological consequences assessment of technological disasters using satellite observations'. Proc. of the 4th International Workshop on Reliable Engineering Computing "Robust Design – Coping with Hazards, Risk and Uncertainty", March 3–5, 2010, Singapore, National University of Singapore, 765–776.

Yu. Kostyuchenko, V. (2015). 'Geostatistics and remote sensing for extremes forecasting and disaster risk multiscale analysis', in Numerical Methods for Reliability and Safety Assessment: Multiscale and Multiphysics Systems, S. Kadry and A. El Hami, (eds.), 404–423, Switzerland: Springer International Publishing, DOI 10.1007/978-3-319-07167-1_16, ISBN 978-3-319-07166-4.

6

Performance Measures of a Complex System with Possible Online Repair

Beena Nailwal[1,*], Bhagawati Prasad Joshi[2] and Suraj Bhan Singh[1]

[1]G. B. Pant University of Agriculture and Technology, Pantnagar
[2]Seemant Institute of Technology, Pithoragarh, India
E-mail: bn4jan@gmail.com; bpjoshi.13march@gmail.com;
drsurajbsingh@yahoo.com
*Corresponding Author

In this chapter, performance measures of a complex system with possibility of online repair are evaluated by using supplementary variable technique, Laplace transformation, and copula methodology. Here reliability model considered consists of a complex system in which two subsystems A and B are arranged in series. Subsystem A is circular consecutive 2-out-of-3: F system and subsystem B is circular consecutive 1-out-of-3: P and 2-out-of-3: F system. The repair of the subsystem is done both at partial failure and complete failure with the help of a server. Here it is assumed that during partial failure of the subsystem B, an inspection is carried out by the server to see the feasibility of online repair. Repair of the subsystem at complete failure is done without inspection. The various transition state probabilities along with the asymptotic behavior of the system and some characteristics such as reliability, availability, mean time to failure (MTTF), cost effectiveness, and sensitivity of the system are obtained. Graphs are also plotted to have a comparison of reliability measures and sensitivity analysis for a particular case.

6.1 Introduction

To deal with real life problems, it is not always promising to have high cost original unit, so many company manufacture such a system in which low cost duplicate unit (substandard) are used in place of original unit. Malik

143

et al. (2010) analyzed a system of two like units in which one is standard unit (called original) and other is substandard unit (called duplicate) with the assumption that on failure of original unit if duplicate unit is not able to execute desired function then system operation stopped and repair of the standard unit is taken up. But it is not always true that the reliability and desirability of duplicate unit may differ from original unit. One can observe in daily life that the duplicate unit system is as efficient as original unit and also operating for same period of time as original unit. Many time common men prefer duplicate unit system due to lower cost and good efficiency. Using the duplicate unit as efficient as original unit is not enough to increase the operational ability of the system but repair strategy adopted during the failure of the system also matters. We can see many systems in real life which might be failed totally either from good state or via partial failure. Electric transformer is a good example of this kind. Also sometime it becomes essential to repair the partially failed unit or subsystem online to enlarge the operational capability of the system. Further, it is not always seen that the repair of the partially failed unit or subsystem is done at down state. In such a situation, inspection plays an important role to decide whether the repair can be done online or at down state. Some authors including (Gopalan and Naidu, 1982; Malik, 2008; Malik and Pawar, 2010; Malik et al., 2010; Pawar and Malik, 2011; Tuteja and Malik, 1994) have analyzed a system with different sets of assumptions on inspection and online repair policy. Furthermore, the system configuration also affects the performance of a system. Researchers (Chiang and Niu, 1981; Mishra and Balagurusamy, 1976) have analyzed system model having different configurations such as k-out-of-m: G and consecutive k-out-of-n: F. Systems having different configurations possess different properties. A circular consecutive k-out-of-n: F is a type of system configuration having a series of n-ordered components arranged along a circle such that the system will be in failure mode if at least k consecutive components in the system fail. Yam et al. (2003) considered such type of system. Researchers (Kumar and Singh, 2016; Manglik and Ram, 2015; Nailwal and Singh, 2011, 2012a,b; Kumar and Singh, 2016; Ram, 2010) have discussed and studies different systems under different assumptions, and estimated their respective reliability parameters.

Keeping the above facts in mind, an attempt is made to develop a reliability model of a complex system in which two subsystems namely A and B are arranged in series. Subsystem A is circular consecutive 2-out-of-3: F system and subsystem B is circular consecutive 1-out-of-3: P and 2-out-of-3: F system. Subsystem A consists of three units A_i ($i = 1, 2, 3$) arranged in a circle and two consecutive units are essential to fail for the failure of the subsystem and subsystem B consists of three units B_j ($j = 1, 2, 3$)

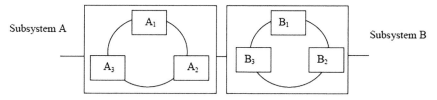

Figure 6.1 Diagram of investigated system.

arranged along a circle and one unit is required to fail for the partial failure of the subsystem and two consecutive units are required to fail for the failure of the subsystem. In both the subsystems A and B, two units are original and one unit is duplicate (substandard unit). Here we assume that duplicate unit is as efficient as original unit, i.e., both the units either original unit or duplicate unit has same failure rate. There is a single server who is always available whenever needed. Server inspects the subsystem B at partial failure to see the possibility of online repair. If online repair is not feasible then the subsystem is repaired at down state. Repair of the system at complete failure is done without inspection. The system is repaired as good as new after partial failure and complete failure. The Gumbel-Hougaard family of copula (Nelson, 2006; Sen, 2003) is applied to estimate joint distribution of repairs whenever both the subsystems are being repaired simultaneously with dissimilar repair rates. In general, it is assumed that failure rates are constant whereas the repairs and inspection follow general distribution in all the cases. The expressions for performance measures such as reliability, availability, mean time to failure (MTTF), cost effectiveness, and sensitivity analysis are derived using supplementary variable technique, Laplace transformation, and copula methodology. Graphs are plotted to represent the performance of reliability measures for a particular case. Figures 6.1 and 6.2 depict the diagram of investigated system and transition state diagram, respectively. Table 6.1 represents the state specification of the system.

6.2 Assumptions

- At first, the system is in perfectly good state.
- The considered system consists of subsystems A and B linked in series.
- Subsystem A is circular consecutive 2-out-of-3: *F* system and subsystem B is circular consecutive 1-out-of-3: *P* and 2-out-of-3: *F* configuration.
- In subsystems A and B two units are standard units (called original units) and one unit is duplicate unit (called substandard unit).

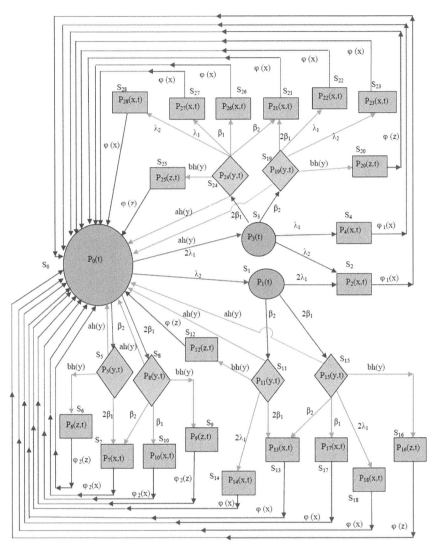

Figure 6.2 Transition diagram.

- Duplicate unit is as efficient as original unit.
- Only a single repairman is available to repair the system instantly whenever required.
- Server inspects the subsystem B at partial failure to observe the likelihood of online repair. If online repair is not possible then the subsystem is repaired at down state.

Table 6.1 State specification

States	Subsystem A: Good Units	Subsystem B: Good Units	System State	States	Subsystem A: Good Units	Subsystem B: Good Units	System State
S_0	A_1, A_2, A_3	B_1, B_2, B_3	W	S_{15}	A_1, A_3	B_1 or B_3, B_2	W_I
S_1	A_1, A_3	B_1, B_2, B_3	W	S_{16}	A_1, A_3	B_1 or B_3, B_2	D_R
S_2	A_1 or A_3	B_1, B_2, B_3	F_R	S_{17}	A_1, A_3	B_2	F_R
S_3	A_1 or A_3, A_2	B_1, B_2, B_3	W	S_{18}	A_1 or A_3	B_1 or B_3, B_2	F_R
S_4	A_2	B_1, B_2, B_3	F_R	S_{19}	A_1 or A_3, A_2	B_1, B_3	W_I
S_5	A_1, A_2, A_3	B_1, B_3	W_I	S_{20}	A_1 or A_3, A_2	$B_1, B_3,$	D_R
S_6	A_1, A_2, A_3	B_1, B_3	D_R	S_{21}	A_1 or A_3, A_2	B_1 or B_3	F_R
S_7	A_1, A_2, A_3	B_1 or B_3	F_R	S_{22}	A_2	B_1, B_3	F_R
S_8	A_1, A_2, A_3	B_1 or B_3, B_2	W_I	S_{23}	A_1 or A_3	B_1, B_3	F_R
S_9	A_1, A_2, A_3	B_1 or B_3, B_2	D_R	S_{24}	A_1 or A_3, A_2	B_1 or B_3, B_2	W_I
S_{10}	A_1, A_2, A_3	B_2	F_R	S_{25}	A_1 or A_3, A_2	B_1 or B_3, B_2	D_R
S_{11}	A_1, A_3	B_1, B_3	W_I	S_{26}	A_1 or A_3, A_2	B_2	F_R
S_{12}	A_1, A_3	B_1, B_3	D_R	S_{27}	A_2	B_1 or B_3, B_2	F_R
S_{13}	A_1, A_3	B_1 or B_3	F_R	S_{28}	A_1 or A_3	B_1 or B_3, B_2	F_R
S_{14}	A_1 or A_3	B_1, B_3	F_R				

Note: W: Working state, W_I: Working under inspection, F_R: Failed under repair, D_R: System is repaired at down state.

- Repair of the subsystem at total failure is carried out without examination.
- Repair of the subsystem is carried out both at partial failure and total failure.
- The joint probability distribution of repairs when both the subsystems A and B are under repair is given by Gumbel-Hougaard family of copula.

6.3 Notations

λ_1 : Failure rate of original units 1 and 3 of subsystem A.
λ_2 : Failure rate of duplicate unit 2 of subsystem A.
μ_1 : Failure rate of original units 1 and 3 of subsystem B.
μ_2 : Failure rate of duplicate unit 2 of subsystem B.
$\varphi_1(x)$: Repair rate of subsystem A.
$\varphi_2(x)$: Repair rate of subsystem B.
$\varphi_1(z)$: Repair rate when some failed units of subsystem A are repaired.
$\varphi_2(z)$: Repair rate of subsystem B at partial failure.

$\varphi(z)$: Coupled repair rate, when some failed units of subsystem A and subsystem B are repaired but system is working.

$h(y)$: Inspection rate at partial failure of subsystem B.

a/b : Change rate of partially failed unit to online repair/repair in down state.

x/y z : Elapsed repair time at complete failed state/Elapsed inspection time at partially failed state/Elapsed repair time at down state.

$P_i(t)$: Probability when system is in S_i state at time t for $i = 1$ to $i = 27$.

$\overline{P}_i(s)$: Laplace transform of $P_i(t)$.

$P_i(x, t)$: Probability density function at instant t when system is in failed state S_i and system is under repair, elapsed repair time is x.

$P_i(y, t)$: Probability density function at instant t when system is in partially failed state S_i and is under inspection elapsed inspection time is y.

$P_i(z, t)$: Probability density function at instant t when system is in down state S_i and system is under repair, elapsed repair time is z.

$E_p(t)$: Expected profit in the interval $(0, t]$.

K_1, K_2: Revenue and service cost per unit time, respectively.

$S_\eta(x)$: $\eta(x) \exp(-\int_0^x \eta(x))\, dx$

$\overline{S}_\eta(x)$: Laplace transform of $S_\eta(x) = \int_0^\infty \eta(x) \exp(-sx - \int_0^x \eta(x))\, dx$

If $u_1 = \varphi_1(x)$, $u_2 = \varphi_2(x)$ then the joint probability as per Gumbel-Hougaard family of copula is given by

$$C_\theta(u_1,\, u_2) = \exp[-\{(-\log \varphi_1(x))^\theta + (-\log \varphi_2(x))^\theta\}^{1/\theta}]$$

6.4 Formulation of Mathematical Model

By elementary probability and continuity arguments, the set of integrodifferential equations corresponding to the system is given below

$$\left[\frac{d}{dt} + \lambda_2 + 2\lambda_1 + \beta_2 + 2\beta_1\right] P_0(t)$$

$$= \int_0^\infty \varphi_2(z) P_6(z, t)\, dz + \int_0^\infty ah(y) P_5(y, t)\, dy + \int_0^\infty \varphi_2(z) P_9(z, t)\, dz$$

$$+ \int_0^\infty \varphi_2(x) P_{10}(x, t)\, dx + \int_0^\infty ah(y) P_8(y, t)\, dy + \int_0^\infty \varphi_2(x) P_7(x, t)\, dx$$

$$+ \int_0^\infty \varphi(x) P_{14}(x, t)\, dx + \int_0^\infty \varphi(x) P_{13}(x, t)\, dx + \int_0^\infty \varphi(x) P_{17}(x, t)\, dx$$

$$+ \int_0^\infty \varphi(x) P_{18}(x,t)\,dx + \int_0^\infty \varphi_1(x) P_2(x,t)\,dx + \int_0^\infty \varphi_1(x) P_4(x,t)\,dx$$

$$+ \int_0^\infty \varphi(x) P_{28}(x,t)\,dx + \int_0^\infty \varphi(x) P_{27}(x,t)\,dx + \int_0^\infty \varphi(x) P_{26}(x,t)\,dx$$

$$+ \int_0^\infty \varphi(z) P_{12}(z,t)\,dz + \int_0^\infty ah(y) P_{11}(y,t)\,dy + \int_0^\infty ah(y) P_{15}(y,t)\,dy$$

$$+ \int_0^\infty \varphi(z) P_{16}(z,t)\,dz + \int_0^\infty ah(y) P_{19}(y,t)\,dy + \int_0^\infty \varphi(z) P_{20}(z,t)\,dz$$

$$+ \int_0^\infty ah(y) P_{24}(y,t)\,dy + \int_0^\infty \varphi(z) P_{25}(z,t)\,dz + \int_0^\infty \varphi(x) P_{21}(x,t)\,dx$$

$$+ \int_0^\infty \varphi(x) P_{22}(x,t)\,dx + \int_0^\infty \varphi(x) P_{23}(x,t)\,dx, \tag{6.1}$$

$$\left[\frac{d}{dt} + 2\lambda_1 + \beta_2 + 2\beta_1 \right] P_1(t) = \lambda_2 P_0(t), \tag{6.2}$$

$$\left[\frac{\partial}{\partial t} + \frac{\partial}{\partial x} + \varphi_1(x) \right] P_2(x,t) = 0, \tag{6.3}$$

$$\left[\frac{d}{dt} + \lambda_1 + \lambda_2 + \beta_2 + 2\beta_1 \right] P_3(t) = 2\lambda_1 P_0(t), \tag{6.4}$$

$$\left[\frac{\partial}{\partial t} + \frac{\partial}{\partial x} + \varphi_1(x) \right] P_4(x,t) = 0 \tag{6.5}$$

$$\left[\frac{\partial}{\partial t} + \frac{\partial}{\partial y} + ah(y) + bh(y) + 2\beta_1 \right] P_5(y,t) = 0, \tag{6.6}$$

$$\left[\frac{\partial}{\partial t} + \frac{\partial}{\partial z} + \varphi_2(z) \right] P_6(z,t) = 0, \tag{6.7}$$

$$\left[\frac{\partial}{\partial t} + \frac{\partial}{\partial x} + \varphi_2(x) \right] P_7(x,t) = 0, \tag{6.8}$$

$$\left[\frac{\partial}{\partial t} + \frac{\partial}{\partial y} + ah(y) + bh(y) + \beta_1 + \beta_2 \right] P_8(y,t) = 0, \tag{6.9}$$

$$\left[\frac{\partial}{\partial t} + \frac{\partial}{\partial z} + \varphi_2(z) \right] P_9(z,t) = 0, \tag{6.10}$$

$$\left[\frac{\partial}{\partial t} + \frac{\partial}{\partial x} + \varphi_2(x) \right] P_{10}(x,t) = 0, \tag{6.11}$$

$$\left[\frac{\partial}{\partial t} + \frac{\partial}{\partial y} + ah(y) + bh(y) + 2\beta_1 + 2\lambda_1 \right] P_{11}(y,t) = 0, \tag{6.12}$$

$$\left[\frac{\partial}{\partial t} + \frac{\partial}{\partial z} + \varphi(z) \right] P_{12}(z, t) = 0, \tag{6.13}$$

$$\left[\frac{\partial}{\partial t} + \frac{\partial}{\partial x} + \varphi(x) \right] P_{13}(x, t) = 0, \tag{6.14}$$

$$\left[\frac{\partial}{\partial t} + \frac{\partial}{\partial x} + \varphi(x) \right] P_{14}(x, t) = 0, \tag{6.15}$$

$$\left[\frac{\partial}{\partial t} + \frac{\partial}{\partial y} + \beta_1 + \beta_2 + 2\lambda_1 + ah(y) + bh(y) \right] P_{15}(y, t) = 0, \tag{6.16}$$

$$\left[\frac{\partial}{\partial t} + \frac{\partial}{\partial z} + \varphi(z) \right] P_{16}(z, t) = 0, \tag{6.17}$$

$$\left[\frac{\partial}{\partial t} + \frac{\partial}{\partial x} + \varphi(x) \right] P_{17}(x, t) = 0, \tag{6.18}$$

$$\left[\frac{\partial}{\partial t} + \frac{\partial}{\partial x} + \varphi(x) \right] P_{18}(x, t) = 0, \tag{6.19}$$

$$\left[\frac{\partial}{\partial t} + \frac{\partial}{\partial y} + ah(y) + bh(y) + \lambda_1 + \lambda_2 + 2\beta_1 \right] P_{19}(y, t) = 0, \tag{6.20}$$

$$\left[\frac{\partial}{\partial t} + \frac{\partial}{\partial z} + \varphi(z) \right] P_{20}(z, t) = 0, \tag{6.21}$$

$$\left[\frac{\partial}{\partial t} + \frac{\partial}{\partial x} + \varphi(x) \right] P_{21}(x, t) = 0, \tag{6.22}$$

$$\left[\frac{\partial}{\partial t} + \frac{\partial}{\partial x} + \varphi(x) \right] P_{22}(x, t) = 0, \tag{6.23}$$

$$\left[\frac{\partial}{\partial t} + \frac{\partial}{\partial x} + \varphi(x) \right] P_{23}(x, t) = 0, \tag{6.24}$$

$$\left[\frac{\partial}{\partial t} + \frac{\partial}{\partial y} + ah(y) + bh(y) + \beta_1 + \beta_2 + \lambda_1 + \lambda_2 \right] P_{24}(y, t) = 0, \tag{6.25}$$

$$\left[\frac{\partial}{\partial t} + \frac{\partial}{\partial z} + \varphi(z) \right] P_{25}(z, t) = 0, \tag{6.26}$$

$$\left[\frac{\partial}{\partial t} + \frac{\partial}{\partial x} + \varphi(x) \right] P_{26}(x, t) = 0, \tag{6.27}$$

$$\left[\frac{\partial}{\partial t} + \frac{\partial}{\partial x} + \varphi(x) \right] P_{27}(x, t) = 0, \tag{6.28}$$

$$\left[\frac{\partial}{\partial t} + \frac{\partial}{\partial x} + \varphi(x) \right] P_{28}(x, t) = 0. \tag{6.29}$$

Boundary conditions:

$$P_2(0,t) = \lambda_2 P_3(t), \tag{6.30}$$

$$P_4(0,t) = \lambda_1 P_3(t), \tag{6.31}$$

$$P_5(0,t) = \beta_2 P_0(t), \tag{6.32}$$

$$P_6(0,t) = \int_0^\infty bh(y)P_5(y,t)\,dy, \tag{6.33}$$

$$P_7(0,t) = \beta_2 P_8(0,t), \tag{6.34}$$

$$P_8(0,t) = 2\beta_1 P_0(t), \tag{6.35}$$

$$P_9(0,t) = \int_0^\infty bh(y)P_8(y,t)\,dy, \tag{6.36}$$

$$P_{10}(0,t) = \beta_1 P_8(0,t), \tag{6.37}$$

$$P_{11}(0,t) = \beta_2 P_1(t), \tag{6.38}$$

$$P_{12}(0,t) = \int_0^\infty bh(y)P_{11}(y,t)\,dy, \tag{6.39}$$

$$P_{13}(0,t) = \beta_2 P_{15}(0,t), \tag{6.40}$$

$$P_{14}(0,t) = 2\lambda_1 P_{11}(0,t), \tag{6.41}$$

$$P_{15}(0,t) = 2\beta_1 P_1(t), \tag{6.42}$$

$$P_{16}(0,t) = b \int_0^\infty h(y)P_{15}(y,t)\,dy, \tag{6.43}$$

$$P_{17}(0,t) = \beta_1 P_{15}(0,t), \tag{6.44}$$

$$P_{18}(0,t) = 2\lambda_1 P_{15}(0,t), \tag{6.45}$$

$$P_{19}(0,t) = \beta_2 P_3(t), \tag{6.46}$$

$$P_{20}(0,t) = \int_0^\infty bh(y)P_{19}(y,t)\,dy, \tag{6.47}$$

$$P_{21}(0,t) = \beta_2 P_{24}(0,t), \tag{6.48}$$

$$P_{22}(0,t) = \lambda_1 P_{19}(0,t), \tag{6.49}$$

$$P_{23}(0,t) = \lambda_2 P_{19}(0,t), \tag{6.50}$$

$$P_{24}(0,t) = 2\beta_1 P_3(t), \tag{6.51}$$

$$P_{24}(0,t) = 2\beta_1 P_3(t), \tag{6.52}$$

$$P_{25}(0,t) = \int_0^\infty bh(y)P_{24}(y,t)\,dy, \tag{6.53}$$

$$P_{26}(0,t) = \beta_1 P_{24}(0,t), \tag{6.54}$$

$$P_{27}(0,t) = \lambda_1 P_{24}(0,t), \tag{6.55}$$

$$P_{28}(0,t) = \lambda_2 P_{24}(0,t), \tag{6.56}$$

$$\varphi(x) = \exp[-\{(-\log\varphi_1(x))^\theta + (-\log\varphi_2(x))^\theta\}^{1/\theta}],$$

$$\varphi(z) = \exp[-\{(-\log\varphi_1(z))^\theta + (-\log\varphi_2(z))^\theta\}^{1/\theta}].$$

Initial condition:

$P_0(t) = 1$ at $t = 0$ and all other probabilities are zero initially.

6.5 Solution of the Model

Taking Laplace transformation of (6.1–6.56) and on further simplification, one can obtain transition state probabilities of the system as:

$$\overline{P}_0(s) = \frac{1}{D(s)}, \tag{6.57}$$

$$\overline{P}_1(s) = \frac{\lambda_2 \overline{P}_0(s)}{B(s)}. \tag{6.58}$$

where $B(s) = (s + 2\lambda_1 + \beta_2 + 2\beta_1)$

$$\overline{P}_2(s) = \frac{\lambda_2 2\lambda_1 \overline{P}_0(s)}{A(s)} \left[\frac{1 - \overline{S}_{\varphi 1}(s)}{s}\right], \tag{6.59}$$

$$\overline{P}_3(s) = \frac{2\lambda_1 \overline{P}_0(s)}{A(s)}. \tag{6.60}$$

Where

$$A(s) = (s + \lambda_1 + \lambda_2 + \beta_2 + 2\beta_1),$$

$$\overline{P}_4(s) = \frac{\lambda_1 2\lambda_1 \overline{P}_0(s)}{A(s)} \left[\frac{1 - \overline{S}_{\varphi 1}(s)}{s}\right], \tag{6.61}$$

$$\overline{P}_5(s) = \beta_2 \overline{P}_0(s) \left[\frac{1 - \overline{S}_h(s + 2\beta_1)}{s + 2\beta_1}\right], \tag{6.62}$$

$$\overline{P}_6(s) = b\beta_2 \overline{P}_0(s)\overline{S}_h(s + 2\beta_1) \left[\frac{1 - \overline{S}_{\varphi 2}(s)}{s}\right], \tag{6.63}$$

$$\overline{P}_7(s) = 2\beta_2\beta_1 \overline{P}_0(s) \left[\frac{1 - \overline{S}_{\varphi 2}(s)}{s}\right], \tag{6.64}$$

$$\overline{P}_8(s) = 2\beta_1 \overline{P}_0(s) \left[\frac{1 - \overline{S}_h(s + \beta_1 + \beta_2)}{s + \beta_1 + \beta_2}\right], \tag{6.65}$$

$$\overline{P}_9(s) = b2\beta_1\overline{P}_0(s)\overline{S}_h(s + \beta_1 + \beta_2)\left[\frac{1 - \overline{S}_{\varphi_2}(s)}{s}\right], \tag{6.66}$$

$$\overline{P}_{10}(s) = 2\beta_1^2\overline{P}_0(s)\left[\frac{1 - \overline{S}_{\varphi_2}(s)}{s}\right], \tag{6.67}$$

$$\overline{P}_{11}(s) = \frac{\beta_2\lambda_2\overline{P}_0(s)}{B(s)}\left[\frac{1 - \overline{S}_h(s + 2\beta_1 + 2\lambda_1)}{s + 2\beta_1 + 2\lambda_1}\right], \tag{6.68}$$

$$\overline{P}_{12}(s) = \frac{b\beta_2\lambda_2\overline{P}_0(s)\overline{S}_h(s + 2\beta_1 + 2\lambda_1)}{B(s)}\left[\frac{1 - \overline{S}_{\varphi}(s)}{s}\right], \tag{6.69}$$

$$\overline{P}_{13}(s) = \beta_2 2\beta_1\frac{\lambda_2\overline{P}_0(s)}{B(s)}\left[\frac{1 - \overline{S}_{\varphi}(s)}{s}\right], \tag{6.70}$$

$$\overline{P}_{14}(s) = \beta_2 2\lambda_1\frac{\lambda_2\overline{P}_0(s)}{B(s)}\left[\frac{1 - \overline{S}_{\varphi}(s)}{s}\right], \tag{6.71}$$

$$\overline{P}_{15}(s) = 2\beta_1\frac{\lambda_2\overline{P}_0(s)}{B(s)}\left[\frac{1 - \overline{S}_h(s + \beta_1 + \beta_2 + 2\lambda_1)}{s + \beta_1 + \beta_2 + 2\lambda_1}\right], \tag{6.72}$$

$$\overline{P}_{16}(s) = 2\beta_1 b\frac{\lambda_2\overline{P}_0(s)}{B(s)}\overline{S}_h(s + \beta_1 + \beta_2 + 2\lambda_1)\left[\frac{1 - \overline{S}_{\varphi}(s)}{s}\right], \tag{6.73}$$

$$\overline{P}_{17}(s) = 2\beta_1^2\frac{\lambda_2\overline{P}_0(s)}{B(s)}\left[\frac{1 - \overline{S}_{\varphi}(s)}{s}\right], \tag{6.74}$$

$$\overline{P}_{18}(s) = 2\lambda_1 2\beta_1\frac{\lambda_2\overline{P}_0(s)}{B(s)}\left[\frac{1 - \overline{S}_{\varphi}(s)}{s}\right], \tag{6.75}$$

$$\overline{P}_{19}(s) = \beta_2\frac{2\lambda_1\overline{P}_0(s)}{A(s)}\left[\frac{1 - \overline{S}_h(s + \lambda_1 + \lambda_2 + 2\beta_1)}{s + \lambda_1 + \lambda_2 + 2\beta_1}\right], \tag{6.76}$$

$$\overline{P}_{20}(s) = b\beta_2\frac{2\lambda_1\overline{P}_0(s)}{A(s)}\overline{S}_h(s + \lambda_1 + \lambda_2 + 2\beta_1)\left[\frac{1 - \overline{S}_{\varphi}(s)}{s}\right], \tag{6.77}$$

$$\overline{P}_{21}(s) = \beta_2 2\beta_1\frac{2\lambda_1\overline{P}_0(s)}{A(s)}\left[\frac{1 - \overline{S}_{\varphi}(s)}{s}\right], \tag{6.78}$$

$$\overline{P}_{22}(s) = \lambda_1\beta_2\frac{2\lambda_1\overline{P}_0(s)}{A(s)}\left[\frac{1 - \overline{S}_{\varphi}(s)}{s}\right], \tag{6.79}$$

$$\overline{P}_{23}(s) = \lambda_2\beta_2\frac{2\lambda_1\overline{P}_0(s)}{A(s)}\left[\frac{1 - \overline{S}_{\varphi}(s)}{s}\right], \tag{6.80}$$

$$\overline{P}_{24}(s) = 2\beta_1\frac{2\lambda_1\overline{P}_0(s)}{A(s)}\left[\frac{1 - \overline{S}_h(s + \beta_1 + \beta_2 + \lambda_1 + \lambda_2)}{s + \beta_1 + \beta_2 + \lambda_1 + \lambda_2}\right], \tag{6.81}$$

$$\overline{P}_{25}(s) = b2\beta_1 \frac{2\lambda_1 \overline{P}_0(s)}{A(s)} \overline{S}_h(s + \beta_1 + \beta_2 + \lambda_1 + \lambda_2) \left[\frac{1 - \overline{S}_\varphi(s)}{s}\right], \quad (6.82)$$

$$\overline{P}_{26}(s) = 2\beta_1^2 \frac{2\lambda_1 \overline{P}_0(s)}{A(s)} \left[\frac{1 - \overline{S}_\varphi(s)}{s}\right], \quad (6.83)$$

$$\overline{P}_{27}(s) = \lambda_1 2\beta_1 \frac{2\lambda_1 \overline{P}_0(s)}{A(s)} \left[\frac{1 - \overline{S}_\varphi(s)}{s}\right], \quad (6.84)$$

$$\overline{P}_{28}(s) = \lambda_2 2\beta_1 \frac{2\lambda_1 \overline{P}_0(s)}{A(s)} \left[\frac{1 - \overline{S}_\varphi(s)}{s}\right]. \quad (6.85)$$

Where

$$\begin{aligned}
D(s) = {} & (s + \lambda_2 + 2\lambda_1 + \beta_2 + 2\beta_1) - b\beta_2 \overline{S}_h(s)\overline{S}_{\varphi 2}(s) - a\beta_2 \overline{S}_h(s + 2\beta_1) \\
& - b2\beta_1 \overline{S}_h(s + \beta_1 + \beta_2)\overline{S}_{\varphi 2}(s) - a2\beta_1 \overline{S}_h(s + \beta_1 + \beta_2) \\
& - 2\beta_1\beta_2 \overline{S}_{\varphi 2}(s) - 2\beta_1^2 \overline{S}_{\varphi 2}(s) - \frac{2\lambda_1\beta_2\lambda_2 \overline{S}_\varphi(s)}{B(s)} - \frac{\beta_2 2\beta_1\lambda_2 \overline{S}_\varphi(s)}{B(s)} \\
& - \frac{2\beta_1^2\lambda_2 \overline{S}_\varphi(s)}{B(s)} - \frac{2\lambda_1 2\beta_1\lambda_2 \overline{S}_\varphi(s)}{B(s)} - \frac{\lambda_2 2\lambda_1 \overline{S}_{\varphi 1}(s)}{A(s)} \\
& - \frac{\lambda_1 2\lambda_1 \overline{S}_{\varphi 1}(s)}{A(s)} - \frac{\lambda_2 2\beta_1 2\lambda_1 \overline{S}_\varphi(s)}{A(s)} - \frac{\lambda_1 2\beta_1 2\lambda_1 \overline{S}_\varphi(s)}{A(s)} \\
& - \frac{\beta_1 2\beta_1 2\lambda_1 \overline{S}_\varphi(s)}{A(s)} - \frac{\beta_2 2\beta_1 2\lambda_1 \overline{S}_\varphi(s)}{A(s)} - \frac{\lambda_1 \beta_2 2\lambda_1 \overline{S}_\varphi(s)}{A(s)} \\
& - \frac{\lambda_2 \beta_2 2\lambda_1 \overline{S}_\varphi(s)}{A(s)} - \frac{b\lambda_2\beta_2 \overline{S}_\varphi(s)\overline{S}_h(s + 2\beta_1 + 2\lambda_1)}{B(s)} \\
& - \frac{a\lambda_2\beta_2 \overline{S}_h(s + 2\beta_1 + 2\lambda_1)}{B(s)} - \frac{a\lambda_2 2\beta_1 \overline{S}_h(s + \beta_1 + \beta_2 + 2\lambda_1)}{B(s)} \\
& - \frac{b\lambda_2 2\beta_1 \overline{S}_\varphi(s)\overline{S}_h(s + \beta_1 + \beta_2 + 2\lambda_1)}{B(s)} \\
& - \frac{b\beta_2 2\lambda_1 \overline{S}_\varphi(s)\overline{S}_h(s + \lambda_1 + \lambda_2 + 2\beta_1)}{A(s)} \\
& - \frac{a\beta_2 2\lambda_1 \overline{S}_h(s + \lambda_1 + \lambda_2 + 2\beta_1)}{A(s)} \\
& - \frac{a2\beta_1 2\lambda_1 \overline{S}_h(s + \lambda_1 + \lambda_2 + \beta_1 + \beta_2)}{A(s)}
\end{aligned}$$

$$-\frac{b2\beta_1 2\lambda_1 \overline{S}_\varphi(s)\overline{S}_h(s+\lambda_1+\lambda_2+\beta_1+\beta_2)}{A(s)},$$

$$\varphi(x) = \exp[-\{(-\log\varphi_1(x))^\theta + (-\log\varphi_2(x))^\theta\}^{1/\theta}],$$
$$\varphi(z) = \exp[-\{(-\log\varphi_1(z))^\theta + (-\log\varphi_2(z))^\theta\}^{1/\theta}].$$

Also up and down state probabilities of the system are presented by

$$\overline{P}_{\rm up}(s)$$
$$= \overline{P}_0(s) + \overline{P}_1(s) + \overline{P}_3(s) + \overline{P}_5(s) + \overline{P}_8(s)$$
$$+ \overline{P}_{11}(s) + \overline{P}_{15}(s) + \overline{P}_{19}(s) + \overline{P}_{24}(s)$$
$$= \overline{P}_0(s)\left[1 + \frac{\lambda_2}{B(s)} + \frac{2\lambda_1}{A(s)} + \beta_2\left\{\frac{1-\overline{S}_h(s+2\beta_1)}{s+2\beta_1}\right\}\right.$$
$$+ 2\beta_1\left\{\frac{1-\overline{S}_h(s+\beta_1+\beta_2)}{s+\beta_1+\beta_2}\right\} + \frac{\beta_2\lambda_2}{B(s)}\left\{\frac{1-\overline{S}_h(s+2\beta_1+2\lambda_1)}{s+2\beta_1+2\lambda_1}\right\}$$
$$+ 2\beta_1\frac{\lambda_2}{B(s)}\left\{\frac{1-\overline{S}_h(s+\beta_1+\beta_2+2\lambda_1)}{s+\beta_1+\beta_2+2\lambda_1}\right\}$$
$$+ \beta_2\frac{2\lambda_1}{A(s)}\left\{\frac{1-\overline{S}_h(s+\lambda_1+\lambda_2+2\beta_1)}{s+\lambda_1+\lambda_2+2\beta_1}\right\}$$
$$+ 2\beta_1\frac{2\lambda_1}{A(s)}\left\{\frac{1-\overline{S}_h(s+\beta_1+\beta_2+\lambda_1+\lambda_2)}{s+\beta_1+\beta_2+\lambda_1+\lambda_2}\right\}\right] \tag{6.86}$$

$$\overline{P}_{\rm down}(s)$$
$$= \overline{P}_2(s) + \overline{P}_4(s) + \overline{P}_6(s) + \overline{P}_7(s) + \overline{P}_9(s) + \overline{P}_{10}(s) + \overline{P}_{12}(s)$$
$$+ \overline{P}_{13}(s) + \overline{P}_{14}(s) + \overline{P}_{16}(s) + \overline{P}_{17}(s) + \overline{P}_{18}(s) + \overline{P}_{20}(s) + \overline{P}_{21}(s)$$
$$+ \overline{P}_{22}(s) + \overline{P}_{23}(s) + \overline{P}_{25}(s) + \overline{P}_{26}(s) + \overline{P}_{27}(s) + \overline{P}_{28}(s)$$
$$= \overline{P}_0(s)\left[\frac{\lambda_2 2\lambda_1}{A(s)}\left\{\frac{1-\overline{S}_{\varphi 1}(s)}{s}\right\} + \frac{\lambda_1 2\lambda_1}{A(s)}\left\{\frac{1-\overline{S}_{\varphi 1}(s)}{s}\right\}\right.$$
$$+ b\beta_2\overline{S}_h(s)\left\{\frac{1-\overline{S}_{\varphi 2}(s)}{s}\right\} + 2\beta_2\beta_1\left\{\frac{1-\overline{S}_{\varphi 2}(s)}{s}\right\}$$
$$+ b2\beta_1\overline{S}_h(s+\beta_1+\beta_2)\left\{\frac{1-\overline{S}_{\varphi 2}(s)}{s}\right\} + 2\beta_1^2\left\{\frac{1-\overline{S}_{\varphi 2}(s)}{s}\right\}$$
$$+ \frac{b\beta_2\lambda_2\overline{S}_h(s+2\beta_1+2\lambda_1)}{B(s)}\left\{\frac{1-\overline{S}_\varphi(s)}{s}\right\} + \beta_2 2\beta_1\frac{\lambda_2}{B(s)}\left\{\frac{1-\overline{S}_\varphi(s)}{s}\right\}$$

$$+\beta_2 2\lambda_1 \frac{\lambda_2}{B(s)}\left\{\frac{1-\overline{S}_\varphi(s)}{s}\right\} + 2\beta_1 b \frac{\lambda_2}{B(s)}\overline{S}_h(s+\beta_1+\beta_2+2\lambda_1)$$

$$\left\{\frac{1-\overline{S}_\varphi(s)}{s}\right\} + 2\beta_1^2 \frac{\lambda_2}{B(s)}\left\{\frac{1-\overline{S}_\varphi(s)}{s}\right\} + 2\lambda_1 2\beta_1 \frac{\lambda_2}{B(s)}\left\{\frac{1-\overline{S}_\varphi(s)}{s}\right\}$$

$$+b\beta_2 \frac{2\lambda_1}{A(s)}\overline{S}_h(s+\lambda_1+\lambda_2+2\beta_1)\left\{\frac{1-\overline{S}_\varphi(s)}{s}\right\}$$

$$+\beta_2 2\beta_1 \frac{2\lambda_1}{A(s)}\left\{\frac{1-\overline{S}_\varphi(s)}{s}\right\} + \lambda_1\beta_2 \frac{2\lambda_1}{A(s)}\left\{\frac{1-\overline{S}_\varphi(s)}{s}\right\}$$

$$+\lambda_2\beta_2 \frac{2\lambda_1}{A(s)}\left\{\frac{1-\overline{S}_\varphi(s)}{s}\right\} + b2\beta_1 \frac{2\lambda_1}{A(s)}\overline{S}_h(s+\beta_1+\beta_2+\lambda_1+\lambda_2)$$

$$\left\{\frac{1-\overline{S}_\varphi(s)}{s}\right\} + 2\beta_1^2 \frac{2\lambda_1}{A(s)}\left\{\frac{1-\overline{S}_\varphi(s)}{s}\right\}$$

$$+\lambda_1 2\beta_1 \frac{2\lambda_1}{A(s)}\left\{\frac{1-\overline{S}_\varphi(s)}{s}\right\} + \lambda_2 2\beta_1 \frac{2\lambda_1}{A(s)}\left\{\frac{1-\overline{S}_\varphi(s)}{s}\right\}\bigg]. \tag{6.87}$$

From (6.87) and (6.88), we have

$$\overline{P}_{up}(s) + \overline{P}_{down}(s) = 1/s.$$

6.6 Asymptotic Behavior

Using Able's lemma

$$\lim_{s\to 0}\{s\overline{F}(s)\} = \lim_{t\to\infty} F(t)$$

in Equations (6.86) and (6.87), we obtain below time independent probabilities of up and down state

$$P_{up} = \frac{1}{D(0)}\left[1 + \frac{\lambda_2}{B(0)} + \frac{2\lambda_1}{A(0)} + \beta_2\left\{\frac{1-\overline{S}_h(2\beta_1)}{2\beta_1}\right\}\right.$$

$$+2\beta_1\left\{\frac{1-\overline{S}_h(\beta_1+\beta_2)}{\beta_1+\beta_2}\right\} + \frac{\beta_2\lambda_2}{B(0)}\left\{\frac{1-\overline{S}_h(2\beta_1+2\lambda_1)}{2\beta_1+2\lambda_1}\right\}$$

$$+2\beta_1\frac{\lambda_2}{B(0)}\left\{\frac{1-\overline{S}_h(\beta_1+\beta_2+2\lambda_1)}{\beta_1+\beta_2+2\lambda_1}\right\}$$

$$+\beta_2\frac{2\lambda_1}{A(0)}\left\{\frac{1-\overline{S}_h(\lambda_1+\lambda_2+2\beta_1)}{\lambda_1+\lambda_2+2\beta_1}\right\}$$

$$+ 2\beta_1 \frac{2\lambda_1}{A(0)} \left\{ \frac{1 - \overline{S}_h(\beta_1 + \beta_2 + \lambda_1 + \lambda_2)}{\beta_1 + \beta_2 + \lambda_1 + \lambda_2} \right\} \Bigg]$$

P_{down}

$$= \frac{1}{D(0)} \left[\frac{\lambda_2 2\lambda_1}{A(0)} \overline{M}_{\varphi_1} + \frac{\lambda_1 2\lambda_1}{A(0)} \overline{M}_{\varphi_1} + b\beta_2 H(0) \overline{M}_{\varphi_2} + 2\beta_2 \beta_1 \overline{M}_{\varphi_2} \right.$$

$$+ b2\beta_1 \overline{S}_h(\beta_1 + \beta_2) \overline{M}_{\varphi_2} + 2\beta_1^2 \overline{M}_{\varphi_2} + \frac{b\beta_2 \lambda_2 \overline{S}_h(2\beta_1 + 2\lambda_1)}{B(0)} \overline{M}_\varphi$$

$$+ \beta_2 2\beta_1 \frac{\lambda_2}{B(0)} \overline{M}_\varphi + \beta_2 2\lambda_1 \frac{\lambda_2}{B(0)} \overline{M}_\varphi + 2\beta_1 b \frac{\lambda_2}{B(0)} \overline{S}_h(\beta_1 + \beta_2 + 2\lambda_1) \overline{M}_\varphi$$

$$+ 2\beta_1^2 \frac{\lambda_2}{B(0)} \overline{M}_\varphi + 2\lambda_1 2\beta_1 \frac{\lambda_2}{B(0)} \overline{M}_\varphi + b\beta_2 \frac{2\lambda_1}{A(0)} \overline{M}_\varphi \overline{S}_h(\lambda_1 + \lambda_2 + 2\beta_1)$$

$$+ \beta_2 2\beta_1 \frac{2\lambda_1}{A(0)} \overline{M}_\varphi + \lambda_1 \beta_2 \frac{2\lambda_1}{A(0)} \overline{M}_\varphi + \lambda_2 \beta_2 \frac{2\lambda_1}{A(0)} \overline{M}_\varphi$$

$$+ b2\beta_1 \frac{2\lambda_1}{A(0)} \overline{S}_h(\beta_1 + \beta_2 + \lambda_1 + \lambda_2) \overline{M}_\varphi + 2\beta_1^2 \frac{2\lambda_1}{A(0)} \overline{M}_\varphi$$

$$+ \lambda_1 2\beta_1 \frac{2\lambda_1}{A(0)} \overline{M}_\varphi + \lambda_2 2\beta_1 \frac{2\lambda_1}{A(0)} \overline{M}_\varphi,$$

where

$$D(0) = \lim_{s \to 0} D(s), \quad B(0) = \lim_{s \to 0} B(s), \quad A(0) = \lim_{s \to 0} A(s),$$

$$H(0) = \lim_{s \to 0} \overline{S}_h(s), \quad \overline{M}_\varphi = \lim_{s \to 0} \left\{ \frac{1 - \overline{S}_\varphi(s)}{s} \right\},$$

$$\overline{M}_{\varphi_1} = \lim_{s \to 0} \left\{ \frac{1 - \overline{S}_{\varphi_1}(s)}{s} \right\}, \quad \overline{M}_{\varphi_2} = \lim_{s \to 0} \left\{ \frac{1 - \overline{S}_{\varphi_2}(s)}{s} \right\},$$

6.7 Particular Cases

6.7.1 When Repair and Inspection Follows Exponential Distribution

In this case, the results can be derived by putting

$$\overline{S}_{\varphi_1}(s) = \frac{\varphi_1(x)}{s + \varphi_1(x)}, \quad \overline{S}_{\varphi_2}(s) = \frac{\varphi_2(x)}{s + \varphi_2(x)},$$

$$\overline{S}_h(s) = \frac{h(y)}{s + h(y)} \tag{6.88}$$

in Equations (6.86) and (6.87), which yield

$$\overline{P}_{up}(s)$$

$$= \frac{1}{D_1(s)} \left[1 + \frac{\lambda_2}{B(s)} + \frac{2\lambda_1}{A(s)} + \beta_2 \frac{1}{s + 2\beta_1 + h(y)} \right.$$

$$+ 2\beta_1 \frac{1}{s + \beta_1 + \beta_2 + h(y)} + \frac{\beta_2 \lambda_2}{B(s)} \frac{1}{s + 2\beta_1 + 2\lambda_1 + h(y)}$$

$$+ 2\beta_1 \frac{\lambda_2}{B(s)} \frac{1}{s + \beta_1 + \beta_2 + 2\lambda_1 + h(y)} + \beta_2 \frac{2\lambda_1}{A(s)}$$

$$\frac{1}{s + \lambda_1 + \lambda_2 + 2\beta_1 + h(y)} + \left. 2\beta_1 \frac{2\lambda_1}{A(s)} \frac{1}{s + \beta_1 + \beta_2 + \lambda_1 + \lambda_2 + h(y)} \right]$$

$$\overline{P}_{down}(s)$$

$$= \frac{1}{D_1(s)} \left[\frac{\lambda_2 2\lambda_1}{A(s)} \frac{1}{s + \varphi_1(x)} + \frac{\lambda_1 2\lambda_1}{A(s)} \frac{1}{s + \varphi_1(x)} + b\beta_2 \frac{h(y)}{s + h(y)} \frac{1}{s + \varphi_2(z)} \right.$$

$$+ 2\beta_2 \beta_1 \frac{1}{s + \varphi_2(x)} + b2\beta_1 \overline{S}_h(s + \beta_1 + \beta_2) \frac{1}{s + \varphi_2(z)} + 2\beta_1^2 \frac{1}{s + \varphi_2(x)}$$

$$+ \frac{b\beta_2 \lambda_2}{B(s)} \frac{h(y)}{s + 2\beta_1 + 2\lambda_1 + h(y)} \frac{1}{s + \varphi(z)} + \beta_2 2\beta_1 \frac{\lambda_2}{B(s)} \frac{1}{s + \varphi(x)}$$

$$+ \beta_2 2\lambda_1 \frac{\lambda_2}{B(s)} \frac{1}{s + \varphi(x)} + 2\beta_1 b \frac{\lambda_2}{B(s)} \frac{1}{s + \varphi(z)} \frac{h(y)}{s + \beta_1 + \beta_2 + 2\lambda_1 + h(y)}$$

$$+ 2\beta_1^2 \frac{\lambda_2}{B(s)} \frac{1}{s + \varphi(x)} + 2\lambda_1 2\beta_1 \frac{\lambda_2}{B(s)} \frac{1}{s + \varphi(x)} + b\beta_2 \frac{2\lambda_1}{A(s)}$$

$$\frac{h(y)}{s + \lambda_1 + \lambda_2 + 2\beta_1 + h(y)} \frac{1}{s + \varphi(z)} + \beta_2 2\beta_1 \frac{2\lambda_1}{A(s)} \frac{1}{s + \varphi(x)}$$

$$+ \lambda_1 \beta_2 \frac{2\lambda_1}{A(s)} \frac{1}{s + \varphi(x)} + \lambda_2 \beta_2 \frac{2\lambda_1}{A(s)} \frac{1}{s + \varphi(x)}$$

$$+ b2\beta_1 \frac{2\lambda_1}{A(s)} \frac{h(y)}{s + \beta_1 + \beta_2 + \lambda_1 + \lambda_2 + h(y)} \frac{1}{s + \varphi(z)} + 2\beta_1^2$$

$$\frac{2\lambda_1}{A(s)} \frac{1}{s + \varphi(x)} + \lambda_1 2\beta_1 \frac{2\lambda_1}{A(s)} \frac{1}{s + \varphi(x)} + \left. \lambda_2 2\beta_1 \frac{2\lambda_1}{A(s)} \frac{1}{s + \varphi(x)} \right].$$

Where

$$D(s) = (s + \lambda_2 + 2\lambda_1 + \beta_2 + 2\beta_1) - b\beta_2 \frac{h(y)}{s + h(y)} \frac{\varphi_2(z)}{s + \varphi_2(z)}$$

$$- a\beta_2 \frac{h(y)}{s + 2\beta_1 + h(y)} - b2\beta_1 \frac{h(y)}{s + \beta_1 + \beta_2 + h(y)} \frac{\varphi_2(z)}{s + \varphi_2(z)}$$

$$-a2\beta_1 \frac{h(y)}{s+\beta_1+\beta_2+h(y)} - 2\beta_1\beta_2 \frac{\varphi_2(x)}{s+\varphi_2(x)} - 2\beta_1^2 \frac{\varphi_2(x)}{s+\varphi_2(x)}$$

$$-\frac{2\lambda_1\beta_2\lambda_2}{B(s)}\frac{\varphi(x)}{s+\varphi(x)} - \frac{\beta_2 2\beta_1\lambda_2}{B(s)}\frac{\varphi(x)}{s+\varphi(x)} - \frac{2\beta_1^2\lambda_2}{B(s)}\frac{\varphi(x)}{s+\varphi(x)}$$

$$-\frac{2\lambda_1 2\beta_1\lambda_2}{B(s)}\frac{\varphi(x)}{s+\varphi(x)} - \frac{\lambda_2 2\lambda_1}{A(s)}\frac{\varphi_1(x)}{s+\varphi_1(x)} - \frac{\lambda_1 2\lambda_1}{A(s)}\frac{\varphi_1(x)}{s+\varphi_1(x)}$$

$$-\frac{\lambda_2 2\beta_1 2\lambda_1}{A(s)}\frac{\varphi(x)}{s+\varphi(x)} - \frac{\lambda_1 2\beta_1 2\lambda_1}{A(s)}\frac{\varphi(x)}{s+\varphi(x)} - \frac{\beta_1 2\beta_1 2\lambda_1}{A(s)}\frac{\varphi(x)}{s+\varphi(x)}$$

$$-\frac{\beta_2 2\beta_1 2\lambda_1}{A(s)}\frac{\varphi(x)}{s+\varphi(x)} - \frac{\lambda_1\beta_2 2\lambda_1}{A(s)}\frac{\varphi(x)}{s+\varphi(x)} - \frac{\lambda_2\beta_2 2\lambda_1}{A(s)}\frac{\varphi(x)}{s+\varphi(x)}$$

$$-\frac{b\lambda_2\beta_2}{B(s)}\frac{\varphi(z)}{s+\varphi(z)}\frac{h(y)}{s+2\beta_1+2\lambda_1+h(y)} - \frac{a\lambda_2\beta_2}{B(s)}\frac{h(y)}{s+2\beta_1+2\lambda_1+h(y)}$$

$$-\frac{a\lambda_2 2\beta_1}{B(s)}\frac{h(y)}{s+\beta_1+\beta_2+2\lambda_1+h(y)} - \frac{b\lambda_2 2\beta_1}{B(s)}\frac{\varphi(z)}{s+\varphi(z)}$$

$$\frac{h(y)}{s+\beta_1+\beta_2+2\lambda_1+h(y)} - \frac{b\beta_2 2\lambda_1}{A(s)}\frac{\varphi(z)}{s+\varphi(z)}$$

$$\frac{h(y)}{s+\lambda_1+\lambda_2+2\beta_1+h(y)} - \frac{a\beta_2 2\lambda_1}{A(s)}\frac{h(y)}{s+\lambda_1+\lambda_2+2\beta_1+h(y)}$$

$$-\frac{a2\beta_1 2\lambda_1}{A(s)}\frac{h(y)}{s+\lambda_1+\lambda_2+\beta_1+\beta_2+h(y)}$$

$$-\frac{b2\beta_1 2\lambda_1}{A(s)}\frac{\varphi(z)}{s+\varphi(z)}\frac{h(y)}{s+\lambda_1+\lambda_2+\beta_1+\beta_2+h(y)}$$

6.8 Numerical Computation

To numerically solve the model, failure rates and some other parameters are kept fixed as

$$\lambda_1 = 0.1, \ \lambda_2 = 0.1, \ \beta_1 = 0.1, \ \beta_2 = 0.1, \ \theta = 1, \ x = y = z = 1 \quad (6.89)$$

in order to obtain the reliability, availability, MTTF, cost analysis, and sensitivity analysis of the system. Also let the repairs are following exponential distribution, i.e., Equation (6.88) holds.

6.8.1 Reliability Analysis

For finding the reliability of the system we take $\varphi_1 = \varphi_2 = \varphi = h = 0$ and the values mentioned in equation (6.89). Now by setting $t = 0, 1, 2, 3, 4, 5, 6,$

7, 8, 9, 10, one can obtain Table 6.2 and correspondingly Figure 6.3, which represents how reliability changes as time increases.

6.8.2 Availability Analysis

To evaluate availability, in addition to all values given in Equation (6.89), we take $\varphi_1 = \varphi_2 = \varphi = h = 1$ in Equation (6.86) and taking inverse Laplace

Table 6.2　Time vs. reliability

Time	Reliability
0	1
1	0.951724677
2	0.844533596
3	0.717474775
4	0.592275773
5	0.479535733
6	0.383220553
7	0.303650964
8	0.23936053
9	0.188180955
10	0.147831972

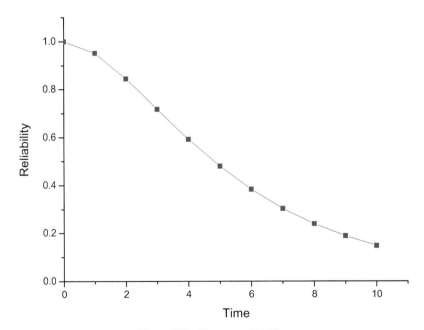

Figure 6.3　Time vs. reliability.

transform when $a = 0.7$ and $b = 0.3$, i.e., if there is more chances of online repair, one can obtain

$$P_{up}(t) = 0.9030556806e^{(-0.01667892203t)} + 0.01474829259e^{(-0.7867471695t)}$$
$$-0.1110165679e^{(-1.132729589t)} + 0.5548012609e^{(-1.179339164t)}$$
$$\sin(0.2724874383t) + 0.0940225505e^{(-1.179339164t)}$$
$$\cos(0.2724874383t) + 0.09919004421e^{(-1.405165992t)} \qquad (6.90)$$

When $a = 0.3$ and $b = 0.7$, i.e., if there is less chances of online repair and system is mostly repaired at down state then availability of the system is given by

$$P_{up}(t)$$
$$= 0.1507905735e^{(-1.196358202t)} \cos(0.4801432397t)$$
$$+0.4490930097e^{(1.196358202t)} \sin(0.4801432397t)$$
$$-0.01416143692e^{(-1.091813083t)} - 0.006288812090e^{(0.7477545156t)}$$
$$+0.8388911508e^{(-0.01266405183t)} + 0.03076852481e^{(-1.455051946t)} \qquad (6.91)$$

Varying t from 0 to 10 in Equations (6.90) and (6.91), one can obtain the behavior of availability with respect to time when (i) $a > b$ and (ii) $a < b$. Table 6.3 and Figure 6.4 are obtained corresponding to availability analysis of the system.

6.8.3 MTTF Analysis

The MTTF of the system can be calculated with the help of the formula given below:

$$\text{MTTF} = \lim_{s \to 0} \overline{P}_{up}(s)$$

With the condition $\varphi_1 = \varphi_2 = \varphi = h = 0$. Variation of MTTF of the system against each failure rate gives following four cases:

(i) Varying λ_1 from 0.10 to 1.0 and assuming all the other values as given in Equation (6.89), we can obtain the variation of MTTF against λ_1.

(ii) Increasing the value of λ_2 from 0.10 to 1.0 and taking the values of other parameters as given in Equation (6.89), we get how MTTF changes against λ_2.

(iii) When $\beta_1 = 0.10, 0.20, 0.30, 0.40, 0.50, 0.60, 0.70, 0.80, 0.90, 1.0$ and all the other parameters have the values as in Equation (6.89), we compute the behavior of MTTF with the growing value of β_1.

Table 6.3 Time vs. availability

Time	Availability $a > b$	Availability $a < b$
0	1	1
1	0.957157945	0.930921567
2	0.905727885	0.858106607
3	0.871760763	0.819648146
4	0.849356556	0.800141766
5	0.832337342	0.787714868
6	0.817550252	0.777401207
7	0.803697026	0.767630119
8	0.790300482	0.758018467
9	0.777196962	0.74850674
10	0.764331891	0.739098767

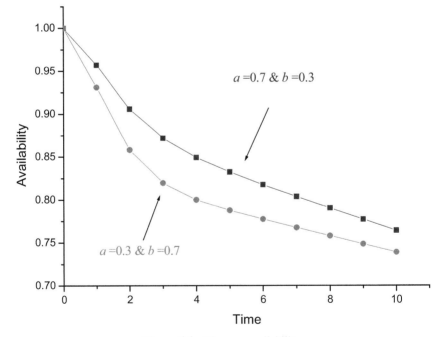

Figure 6.4 Time vs. availability.

(iv) Increase the value of β_2 from 0.10 to 1.0 and assume the other parameters have values as given in Equation (6.89); we obtain the manner in which MTTF varies against β_2.

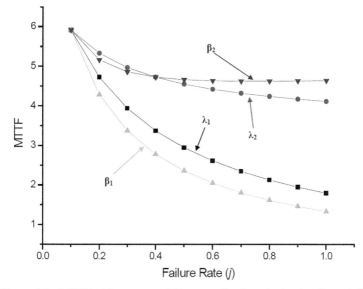

Figure 6.5 MTTF with respect to failure rate (j) where $j = \lambda_1$, λ_2, β_1, and β_2.

Discrepancy of MTTF against λ_1, λ_2, β_1, and β_2 in the above four cases (i), (ii), (iii), and (iv) have been given in Figure 6.5.

6.8.4 Cost Analysis

If service facility is available all the time, then expected profit in $(0, t]$ is given by

$$E_P(t) = K_1 \int_0^t P_{up}(t)\, dt - K_2 t$$

where K_1 and K_2 are the revenue and service cost per unit time, respectively, then by using the expression of $P_{up}(t)$ given in Equation (6.86), one can obtain

$$
\begin{aligned}
E_P(t) \\
&= K_1[-54.14352792e^{(-0.01667892203t)} - 0.01874591122e^{(0.7867471695t)} \\
&\quad +0.09800800560e^{(-1.132729589t)}0.1788701053e^{(-1.179339164t)} \\
&\quad \cos(0.2724874383t)0.4291059091e^{(1.179339164t)}\sin(0.2724874383t) \\
&\quad -0.07058955652e^{(-1.405165992t)} + 54.31372549] - tK_2 \qquad (6.92)
\end{aligned}
$$

This is the equation of expected profit when the possibility of online repair is more, i.e., $a = 0.7$, $b = 0.3$ but if there is possibility of repair at down repair is more, i.e., $a = 0.3$, $b = 0.7$ then expected profit is estimated by

$$E_P(t) = K_1[-0.2383114785e^{(-1.196358202t)}\cos(0.4801432397t)$$
$$-0.2797401011e^{(1.196358202t)}\sin(0.4801432397t)$$
$$+0.01297056899e^{(-1.091813083t)} + 0.008410262939e^{(-0.7477545156t)}$$
$$-66.24192336e^{(-0.01266405183t)} - 0.02114599750e^{(-1.455051946t)}$$
$$+66.48000001] - tK_2 \tag{6.93}$$

Keeping $K_1 = 1$ and varying K_2 as 0.1, 0.2, 0.3, 0.4, 0.5 in Equations (6.92) and (6.93), one can obtain Table 6.5 and correspondingly Figure 6.6.

6.8.5 Sensitivity Analysis

Putting $\varphi_1 = \varphi_2 = \varphi = h = 0$, $\theta = 1$ and $x = y = z = 1$ in Equation (6.86), we have

$$R(s) = \frac{1}{(s + \lambda_2 + 2\lambda_1 + \beta_2 + 2\beta_1)}\left[1 + \frac{\lambda_2}{(s + 2\lambda_1 + \beta_2 + 2\beta_1)}\right.$$
$$+ \frac{2\lambda_1}{(s + \lambda_1 + \lambda_2 + \beta_2 + 2\beta_1)} + \frac{\beta_2}{(s + 2\beta_1)} + \frac{2\beta_1}{(s + \beta_1 + \beta_2)}$$
$$+ \frac{\beta_2\lambda_2}{(s + 2\lambda_1 + \beta_2 + 2\beta_1)(s + 2\lambda_1 + 2\beta_1)}$$

Table 6.4 MTTF with respect to failure rate (j) where $j = \lambda_1, \lambda_2, \beta_1$, and β_2

Failure Rate (P)	w. r. t. λ_1	w. r. t. λ_2	w. r. t. β_1	w. r. t. β_2
0.1	5.91666667	5.91666667	5.91666667	5.91666667
0.2	4.726190476	5.333333335	4.282738097	5.16666667
0.3	3.938492064	4.973214285	3.375	4.866071426
0.4	3.372294372	4.73015873	2.780573594	4.73015873
0.5	2.945665446	4.555555555	2.36080586	4.666666669
0.6	2.613095238	4.424242423	2.049107143	4.638528139
0.7	2.346887998	4.321969698	1.808881886	4.628787879
0.8	2.129186603	4.24009324	1.618304758	4.628982128
0.9	1.947968698	4.173076925	1.463564214	4.634615388
1	1.794853697	4.117216119	1.335505978	4.643190146

$$+\frac{2\beta_1\lambda_2}{(s+2\lambda_1+\beta_2+2\beta_1)(s+2\lambda_1+\beta_2+\beta_1)}$$
$$+\frac{2\beta_2\lambda_1}{(s+\lambda_2+\lambda_1+\beta_2+2\beta_1)(s+\lambda_2+\lambda_1+2\beta_1)}$$
$$\left.+\frac{4\beta_1\lambda_1}{(s+\lambda_2+\lambda_1+\beta_2+2\beta_1)(s+\lambda_2+\lambda_1+\beta_2+\beta_1)}\right] \quad (6.94)$$

Table 6.5 Time vs. expected profit

| | $E_p(t)E_p(t)$ | | | |
| | $a = 0.7, b = 0.3$ | | $a = 0.3, b = 0.7$ | |
Time	$K_2 = 0.2$	$K_2 = 0.4$	$K_2 = 0.2$	$K_2 = 0.4$
0	0	0	0	0
1	0.78300513	0.58300513	0.77212305	0.57212305
2	1.51306099	1.11306099	1.4635541	1.0635541
3	2.20051089	1.60051089	2.10013908	1.50013908
4	2.86041384	2.06041384	2.70908	1.90908
5	3.50098229	2.50098229	3.30270238	2.30270238
6	4.12581269	2.92581269	3.88517916	2.68517916
7	4.73638571	3.33638571	4.45767423	3.05767423
8	5.33335586	3.73335586	5.02048965	3.42048965
9	5.9170833	4.1170833	5.573744	3.773744
10	6.48782893	4.48782893	6.11753772	4.11753772

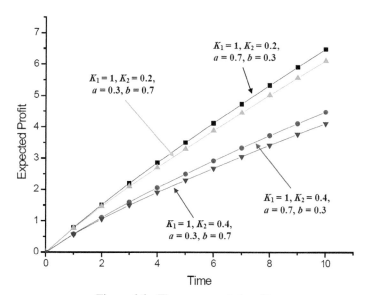

Figure 6.6 Time vs. expected profit.

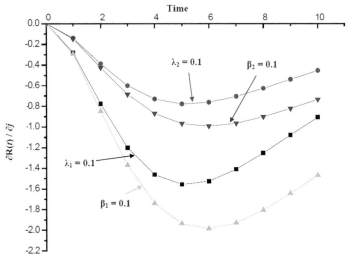

Figure 6.7   Sensitivity of system reliability with respect to j where j denotes different failure rates as λ_1, λ_2, β_1, and β_2.

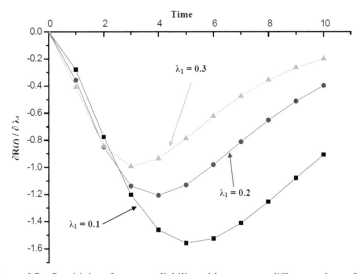

Figure 6.8   Sensitivity of system reliability with respect to different values of λ_1.

We find $\partial R(s)/\partial \lambda_1$ by differentiating Equation (6.94) with respect to λ_1, take its inverse Laplace transform, put $\lambda_1 = 0.1$, $\lambda_2 = 0.1$, $\beta_1 = 0.1$, $\beta_2 = 0.1$ and then varying the time as $t = 0, 1, 2, 3, 4, 5, 6, 7, 8, 9, 10$, we obtain the sensitivity analysis of system reliability with respect to λ_1.

Table 6.6   Sensitivity of system reliability with respect to j where j denotes different failure rates as λ_1, λ_2, β_1, and β_2

Time	Value of $\partial R(t)/\partial j$			
0	0	0	0	0
1	−0.280086266	−0.140043131	−0.293553294	−0.14677665
2	−0.77643659	−0.388218295	−0.849639959	−0.424819979
3	−1.201753245	−0.600876623	−1.371140337	−0.685570168
4	−1.461951535	−0.730975768	−1.739677222	−0.869838611
5	−1.557555424	−0.778777713	−1.936006305	−0.968003154
6	−1.525966081	−0.762983041	−1.986039966	−0.993019984
7	−1.411648705	−0.705824353	−1.929724888	−0.964862444
8	−1.253081357	−0.626540678	−1.805625982	−0.902812993
9	−1.07873209	−0.539366045	−1.64484515	−0.822422575
10	−0.90731759	−0.453658795	−1.469835161	−0.73491758

Table 6.7   Sensitivity of system reliability with respect to different values of λ_1

Time	$\partial R(t)/\partial\lambda_1$	$\partial R(t)/\partial\lambda_1$	$\partial R(t)/\partial\lambda_1$
0	0	0	0
1	−0.280086266	−0.35728606	−0.409046075
2	−0.77643659	−0.849610527	−0.844875639
3	−1.201753245	−1.137237262	−0.993420993
4	−1.461951535	−1.206086753	−0.936073554
5	−1.557555424	−1.129450623	−0.788064137
6	−1.525966081	−0.981035726	−0.623051913
7	−1.411648705	−0.811991833	−0.475589824
8	−1.253081357	−0.651213469	−0.356640433
9	−1.07873209	−0.511777319	−0.265810944
10	−0.90731759	−0.397280804	−0.19847275

6.9 Conclusions

By critical examination of Figure 6.3, we conclude that reliability of the system declines as time inclines and attains a value 0.14 at $t = 10$.

Figure 6.4 shows the availability of the system when (i) possibility of online repair is more than the possibility of repair at down state, i.e., $a > b$ and (ii) when $a < b$. It is clear from Figure 6.4 that availability of the system declines in both the cases as the time increases but the availability of the system is better when there are more chances of online repair. Initially, at time $t = 0$ both reliability and availability of the system has value 1.

Figure 6.5 depicts the behavior of MTTF with respect to λ_1, λ_2, β_1, and β_2. Observation of Figure 6.5 reveals that in each case MTTF of the system decreases as the failure rates increases. MTTF varies from 5.91 to 1.79, 5.91 to 4.11, 5.91 to 1.33, and 5.91 to 4.63 with respect to λ_1, λ_2, β_1, and β_2, respectively.

For the cost analysis of the system we keep revenue cost per unit time at 1 and service cost are kept fixed at 0.2 and 0.4. The behavior of expected profit can be observed from Figure 6.6. The profit goes on decreasing with the inclination of service cost. However, the system becomes more profitable when the chances of online repair are greater than the chances of repair at down state. The maximum and minimum values of expected profit are obtained, and to be 6.487 and 0.583 when the possibility of online repair (a) is greater than the possibility of repair at down state (b) and 6.117 and 0.572 when the possibility of online repair (a) is less than the possibility of repair at down state (b), i.e., $a < b$.

Figure 6.7 represents how sensitivity of system reliability varies regarding to the parameters λ_1, λ_2, β_1, and β_2. From Figure 6.7, it is clear that sensitivity of system reliability initially decreases and then increases as time passes with regard to λ_1, λ_2, β_1, and β_2 at 0.1. It is interesting to note that system reliability is more sensitive respecting to λ_2. Figure 6.8 represents how system reliability varies with the increase of parameter λ_1 from 0.1 to 0.3. The study of the graph reveals that the sensitivity of the system reliability increases with inclination of failure rate λ_1. From Figure 6.8, we can conclude that the system can be prepared less sensitive by controlling its failure rates. The study also concludes that the system becomes more profitable when the possibility of online repair is more.

References

Chiang, D. T., and Niu, S. C. (1981). Reliability of consecutive k-out-of-n: F system, IEEE Transactions on Reliability, R-30, 87–89.

Gopalan, M, N., and Naidu, R. S. (1982). Cost benefit analysis of a one server system subject to inspection, Microelectronics and Reliability, 22, 699–705.

Kumar, D., and Singh, S. B. (2016). Stochastic analysis of complex repairable system with deliberate failure emphasizing reboot delay, Communications in Statistics—Simulation and Computation, 45(2), 583–602.

Malik, S. C., Bhardwaj, R. K., and Grewal, A. S. (2010). Probabilistic analysis of a system of two non-identical parallel units with priority to repair subject to inspection, Journal of Reliability and Statistical Studies, 3(1), 1–11.

Malik, S. C., and Pawar, D. (2010). Reliability and economic measures of a system with inspection for on-line repair and no repair activity in abnormal weather, Bulletin of Pure and Applied Sciences, 29 E(2), 355–368.

Malik, S. C., Pawar, D., and Kumar, J. (2010). Analysis of a system working under different weather conditions with on-line repair and random appearance of the server at partial failure stage, International Journal of Statistics and Systems, 5(4), 485–496.

Malik, S. C. (2008). Reliability modeling and profit analysis of a single-unit system with inspection by a server who appears and disappears randomly, Pure and Applied Mathematika Sciences, LXVII(1-2), 135–146.

Manglik, M., and Ram, M. (2015). Behavioural analysis of a hydroelectric production power plant under reworking scheme, International Journal of Production Research, 53(2), 648–664.

Mishra, K. B., and Balagurusamy, E. (1976). Reliability analysis of k-out-of-m: G system with dependent failures, International Journal of Systems Science, 7(8), 853–861.

Nailwal, B., and Singh, S. B. (2011). Performance evaluation and reliability analysis of a complex system with three possibilities in repair with the application of copula, International Journal of Reliability and Application, 12(1), 15–39.

Nailwal, B., Singh, S. B. (2012). Reliability measures and sensitivity analysis of a complex matrix system including power failure, International Journal of Engineering, 25(1), 115–130.

Nailwal, B., and Singh, S. B. (2012). Reliability and sensitivity analysis of an operating system with inspection in different weather conditions, International Journal of Reliability, Quality and Safety Engineering, 19(2). DOI: 10.1142/S021853931250009X

Nailwal, B., and Singh, S. B. (2017) Reliability analysis of two dissimilar-cold standby redundant systems subject to inspection with preventive maintenance using copula, IGI Global, 201–221. DOI: 10.4018/978-1-5225-1639-2.ch010

Nelson, R. B. (2006). An Introduction to Copulas, 2nd edition, Springer-Verlag: New York.

Pawar, D., and Malik, S. C. (2011). Performance measures of a single—unit system subject to different failure modes with operation in abnormal weather, International Journal of Engineering Science and Technology (IJEST), 3(5), 4084–4089.

Ram, M. (2010). Analysis of a complex system with common cause failure and two types of repair facilities with different distributions in failure, International Journal of Reliability and Safety, 4(4), 381–392.

Sen, P. K. (2003). Copulas: Concepts and novel applications, International Journal of Statistics, LXI(3), 323–353.

Tuteja RK, Malik SC (1994) A system with pre-inspection and two types of repairman, Microelectronics and Reliability, 32(3), 373–377.

Yam, Richard, C. M., Zuo, M. Z., and Zhang, Y. L. (2003). A method for evaluation of reliability indices for repairable circular consecutive- k-out-of-n: F systems, Reliability Engineering and System Safety, 79, 1–9.

7

Small Quadrotor Functioning under Rework Analysis

Nupur Goyal[1], Ayush Kumar Dua[2], Akshay Bhardwaj[2], Akshay Kumar[3,*] and Mangey Ram[4]

[1]Department of Mathematics, Garg P. G. College, Laksar, Uttarakhand, India
[2]Department of Computer Science and Engineering, Graphic Era University, Dehradun 248002, India
[3]Tula's Institute, The Engineering and Management College, Uttarakhand, India
[4]Department of Mathematics, Graphic Era University, Dehradun, India
E-mail: nupurgoyalgeu@gmail.com; duaayush5@gmail.com; akshaykr1001@gmail.com; mangeyram@gmail.com
*Corresponding Author

In present years, reliability has become an important anxiety due to high-tech industrial process with an ever-growing level of sophistication comprises of most engineering systems today. With this rapidly mounting field, this research provides the synopsis of a multicomponent repairable system. This study seeks the several economic characteristics of the reliability of a repairable manmade vertical takeoff and landing (VTOL) prototype system. These characteristics have a great importance in the performance of the VTOL system. This system occupies mainly five independent units and one dependent unit (standby), some independent unit follows k-out-of-n: F redundancy. By a systematic view, this novel work introduces a Markov model of a complex VTOL system from which authors evaluate the system reliability, availability, and mean time to failure by using supplementary variable technique, Laplace and inverse Laplace transformation. They also investigate the system is how sensitive with respect to each failure. Finally, some numerical computation and their graphical presentation are also appended.

7.1 Introduction

The reliability theory tells the assuming of the probability theory and the binary states of an element or subsystems of the system as working or failed. Recently, the stochastic process has a lot of growth to model the basic work in reliability theory. Due to the greater requirement of the highly reliable system than that of the units used, the concept of redundancy and maintenance are often incorporated. So to render more the efficiency of a system, a unit of the system which has failed is renewed by replacing or repaired. In previous years, many researchers and practitioners have paid attention in remodeling the reliability of their pairable system (Kumar et al., 2009). Avizicnis ct al. (2004) introduced the basic concept of the dependability and secure computing and give a precise definition. The systematic careful discussion of dependability and security provides a very easy means of considering several concerns within a single conceptual framework.

Many authors including, Hoffmann et al. (2007) instructed many problems discovered in quadrotor aircraft operating at higher speeds and in the presence of wind disturbances that arise when deviating pointedly from the hover flight regime. They validated the results with thrust test stand measurements and vehicle flight test using Stanford Test bed of Autonomous Rotocraft for Multi-Agent Control (STARMAC) quadrotor helicopter. Bouabdallah and Siegwart (2007) discussed the findings of OS4 project and described the control approach that permitted the design of the main controllers. Berbra et al. (2008) presented the impact of an NCS network system and developed an indicator which is sensitive to packet losses. They found the result using a simulation of a four-rotor helicopter. Sa and Corke (2011) developed a velocity estimator for determining the dynamics of microcopter and shown by closed-loop position control using a nested control structure. Cui and Li (2007), Ram and Singh (2010), and Manglik and Ram (2013) have discussed the reliability measures of many complex systems but they did not consider the vertical takeoff and landing (VTOL) quadrotor system. Quadcopters are compact aircraft vehicles with VTOL. It is a key challenge for civilians to develop a cheap and vigorous system that can achieve its task with required safety level. While a lot of work has been done on a quadrotor, but no one finds out the reliability indices of the quadrotor. Olson and Atkins (2013) analyzed the failure of the Michigan Autonomous Aerial vehicles team's quadrotor unmanned aircraft system and compute cost constraints, weight, and volume. Goyal et al. (2015) studied a computer system based on home or office and calculate reliability, cost analysis, and mean time to failure (MTTF) with the help of copula technique.

Ram and Kumar (2015) consider a 1-out-of-2: *G* system based on *k*-out-of-*n* redundancy system and computed the reliability, sensitivity, MTTF, of the proposed system based on supplementary technique. Goyal et al. (2015) evaluated the reliability measures such as availability, sensitivity, MTTF of the Automated Teller Machine system by using Markov chain process. Ram et al. (2015), Ram and Goyal (2015) presented an engineering model based on semi-Markov power plant and fault tolerant system and evaluate system availability, cost and sensitivity using supplementary technique with the help of Markov process. Goyal et al. (2016) considered a thermal power plant to evaluate the reliability of the proposed system with the help of Markov chain process. Goyal et al. (2016) discussed the probability of water cooling system and evaluate the reliability and cost of the proposed system using supplementary technique. Goyal et al. (2017) proposed a solar thermal power plant system and compute the system reliability of each state by using Markov chain process. Ram and Davim (2017) evaluated the reliability and optimization of multistate system in the engineering field by using different techniques. Goyal and Ram (2017) considered a series-parallel complex system in which two subsystems connected in series manner and computed system reliability with the help of Markov chain technique.

In this study, the considered complex system uses *k*-out-of-*n* redundancy in the series configuration of some components. In the designed VTOL system, four motors are connected in the series system, i.e., 1-out-of-4: *F*. Similarly, four motor controllers used 1-out-of-4: *F* redundancy and four propellers also used the same pattern. When the power supply through the battery has been failed, the standby unit will work. So in contrast to the earlier models, authors have found the availability, reliability, MTTF, and sensitivity of reliability and MTTF with the help of supplementary variable technique, Markov process and the Laplace transformation regarding the various failure rates.

7.2 Mathematical Description of the Propose Model

7.2.1 Assumptions

The following assumptions are used throughout this work:

    i. Initially the system is free from failures.
    ii. At any time *t*, the system covers any one of the three states namely good, degraded, and completely failed state.
    iii. The repair facility is available always.

  iv. Failure and repair rates have general distribution.
  v. After repair, the system works like a new one.
  vi. If the power supply of the system is failed, then the system working with standby.
  vii. The system follows *k*-out-of-*n* type redundancy.

7.2.2 Nomenclature

Notations related with this work are shown in Table 7.1.

7.2.3 Model Formulation

In this research work, authors are considering an aircraft VTOL system, similar to a helicopter; the difference is that it occupies four rotor disks while helicopter occupies three rotor disks. It is a prototype model; it is an unconventional efficient multirotor design. The beauty of that four ordinary brushless out runners directly driving ordinary propellers can be used. It is constructed by some parts and materials, but for our research, we are taking its main parts such as four motors, four motor controllers, flight controllers, four propellers, battery, and one standby unit which work when the power supply through battery has been failed. The configuration diagram of this VTOL system is demonstrated by Figure 7.1(a). On the basis of these parts, authors design the state transition diagram shown in Figure 7.1(b) that contains total seven states in which one is good, one is partially failed (degraded) and

Table 7.1 Notations

T	Time scale.
S	Laplace transform variable.
$\bar{P}(s)$	Laplace transformation of $P(t)$.
$\lambda_{PS}/\lambda_M/\lambda_{MC}/$ $\lambda_F/\lambda_{SPS}/\lambda_P$	Failure rate for power supply/motor/motor controller/flight controller/standby/propeller.
$P_i(t)$	The probability of the stage S_i at time t when $i = 0, 1$.
$P_j(x,t)$	The probability density function of the state S_j, when $j = 2, 3, 4, 5, 6$.
$\mu(x)$	Repair rate for the complete failed state of the system.
$\phi(y)$	Repair rate for the degraded state of the system.
S_0	Good working state of the system.
S_1	Partially failed (degraded) state of the system when the power supply of the system has been failed.
S_j	Completely failed state of the system when the system has been failed by any other failure except the power supply.
$P_{up}(t)$	Up state system probability at time t.
$R(t)$	The reliability of the system at time t.

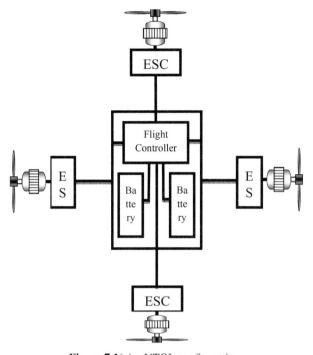

Figure 7.1(a)   VTOL configuration.

five are a completely failed state. Motors are connected in the series system. Motor controllers are connected in series with each other. Propellers are also connected in the series system.

7.2.4 Illustration of the Proposed Model

By probability examination and continuity arguments, the following differential equations overriding the behavior of the system may appear to be good.

$$\left[\frac{\partial}{\partial t} + \lambda_{PS} + \lambda_P + \lambda_M + \lambda_{MC} + \lambda_F\right] P_0(t)$$

$$= \phi(y)P_1(t) + \int_0^\infty \mu(x) \sum_{i=2}^{6} P_i(x,t)\,dx, \qquad (7.1)$$

$$\left[\frac{\partial}{\partial t} + \lambda_{SPS} + \lambda_P + \lambda_M + \lambda_{MC} + \lambda_F + \phi(y)\right] P_1(t) = \lambda_{PS}P_0(t), \quad (7.2)$$

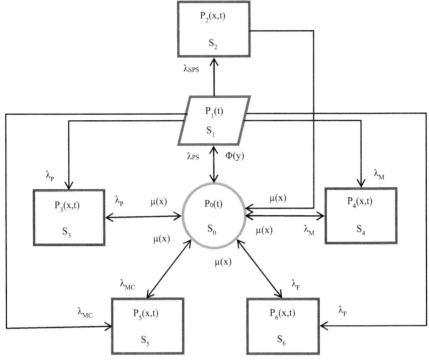

Figure 7.1(b)   State transition diagram.

$$\left[\frac{\partial}{\partial t} + \frac{\partial}{\partial x} + \mu(x)\right] P_i(t) = 0, \quad i = 2, 3, 4, 5, 6. \tag{7.3}$$

Boundary conditions:

$$P_2(0, t) = \lambda_{SPS} P_1(t), \tag{7.4}$$

$$P_i(0, t) = \alpha\beta, \quad i = 3, 4, 5, 6;$$
$$\alpha = \lambda_P, \lambda_M, \lambda_{MC}, \lambda_F; \quad \beta = P_0(t) + P_1(t) \tag{7.5}$$

Initial condition:

$$P_i(0) = 1 \quad if \ i = 0 \ or \ 0 \ if \ i \geq 1. \tag{7.6}$$

Now, taking Laplace transformation of Equations (7.1)–(7.5) using (7.6)

$$[s + \lambda_{PS} + \lambda_P + \lambda_M + \lambda_{MC} + \lambda_F]\overline{P}_0(s)$$

$$= 1 + \phi(y)\overline{P}_1(s) + \int_0^\infty \sum_{i=2}^6 \mu(x)\overline{P}_i(x, s)\, dx, \tag{7.7}$$

$$[s + \lambda_{SPS} + \lambda_P + \lambda_M + \lambda_{MC} + \lambda_F + \phi(y)]\,\overline{P}_1(s) = \lambda_{PS}\overline{P}_0(s), \quad (7.8)$$

$$\left[s + \frac{\partial}{\partial x} + \mu(x)\right]\overline{P}_i(s) = 0, \quad i = 2, 3, 4, 5, 6, \quad (7.9)$$

$$\overline{P}_2(0, s) = \lambda_{SPS}\overline{P}_1(s), \quad (7.10)$$

$$\overline{P}_i(0, s) = \alpha\beta, \quad i = 3, 4, 5, 6;$$

$$\alpha = \lambda_P, \lambda_M, \lambda_{MC}, \lambda_F; \quad \beta = \overline{P}_0(s) + \overline{P}_1(s). \quad (7.11)$$

After solving Equations (7.7)–(7.10) we have the Laplace transformation of transition state probabilities as

$$\overline{P}_0(s) = \frac{s + A_2}{D(s)}, \quad (7.12)$$

$$\overline{P}_1(s) = \frac{\lambda_{PS}}{s + A_2}\overline{P}_0(s), \quad (7.13)$$

$$\overline{P}_2(s) = \left(\frac{1 - S_\mu(s)}{s}\right)\frac{\lambda_{SPS}\lambda_{PS}}{s + A_2}\overline{P}_0(s), \quad (7.14)$$

$$\overline{P}_i(s) = \left(\frac{1 - S_\mu(s)}{s}\right)\left\{\frac{s + A_2 + \lambda_{PS}}{s + A_2}\right\}\alpha\overline{P}_0(s); \quad i = 3, 4, 5, 6;$$

$$\alpha = \lambda_P, \lambda_M, \lambda_{MC}, \lambda_F; \quad (7.15)$$

where

$$D(s) = (s + A_1)(s + A_2) - \phi(y)\lambda_{PS} - A_3\overline{S}_\mu(s),$$

$$C_1 = \lambda_P + \lambda_M + \lambda_{MC} + \lambda_F, \quad A_1 = \lambda_{PS} + C_1,$$

$$A_2 = \lambda_{SPS} + C_1 + \phi(y), \quad A_3 = \lambda_{SPS}\lambda_{PS} + (s + \lambda_{PS} + A_2)C_1.$$

Laplace transformation of the probability of the system is in up-state

$$\overline{P}_{up}(s) = \overline{P}_0(s) + \overline{P}_1(s)$$

$$= \left\{\frac{s + A_2 + \lambda_{PS}}{s + A_2}\right\}\overline{P}_0(s), \quad (7.16)$$

$$\overline{P}_{down}(s) = \sum_{i=2}^{6}\overline{P}_i(s)$$

$$= \left(\frac{1 - S_\mu(s)}{s}\right)\frac{A_3}{s + A_2}\overline{P}_0(s). \quad (7.17)$$

7.3 Particular Cases and Numerical Appraisal

7.3.1 Availability Analysis

Availability is the probability that a system or any part of the system is performing its required function over a period of time in which system is operational. Assuming the values of failure rates and repair rates as $\lambda_{PS} = 0.045$, $\lambda_M = 0.012$, $\lambda_{MC} = 0.024$, $\lambda_F = 0.024$, $\lambda_{SPS} = 0.045$, $\lambda_P = 0.007$, $\mu(x) = 1$, $\phi(y) = 1$ in Equation (7.16) and taking inverse Laplace transformation, one obtain availability in terms of time as

$$P_{up}(t) = 0.9356723254 + (0.004532374101t + 0.06432767455)e^{(-1.112t)}.$$
(7.18)

Now, changing the time unit t from 0 to 20 in Equation (7.18), and obtain the availability of the designed system which is shown in Table 7.2 and determined by the graph in Figure 7.2.

Table 7.2   Availability as function of time

Time (t)	Availability $P_{up}(t)$
0	1.00000
1	0.95823
2	0.94334
3	0.93795
4	0.93590
5	0.93502
6	0.93455
7	0.93422
8	0.93394
9	0.93368
10	0.93342
11	0.93317
12	0.93292
13	0.93267
14	0.93242
15	0.93216
16	0.93191
17	0.93166
18	0.93141
19	0.93116
20	0.93091

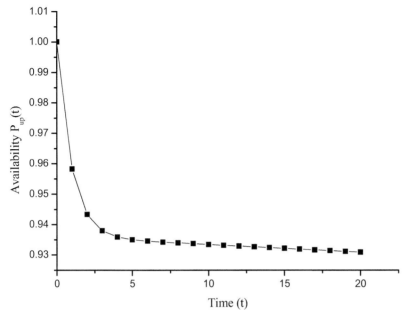

Figure 7.2 Availability as a function of time.

7.3.2 Reliability Analysis

Now putting all repair rates zero in Equation (7.16), we get

$$\overline{R}(s)$$
$$= \frac{s + \lambda_{PS} + \lambda_M + \lambda_{MC} + \lambda_F + \lambda_{SPS} + \lambda_P}{(s + \lambda_{PS} + \lambda_M + \lambda_{MC} + \lambda_F + \lambda_P)(s + \lambda_M + \lambda_{MC} + \lambda_F + \lambda_{SPS} + \lambda_P)}.$$
(7.19)

One can determine the reliability of the system in terms of time t by substituting the value of failure rates as $\lambda_{PS} = 0.045$, $\lambda_M = 0.012$, $\lambda_{MC} = 0.024$, $\lambda_F = 0.024$, $\lambda_{SPS} = 0.045$, $\lambda_P = 0.007$, and taking inverse Laplace transformation of condition (7.19). We have

$$R(t) = (0.045\,t + 1)e^{(-0.112\,t)}.$$
(7.20)

Now taking the time unit t from 0 to 20 in condition (7.20) and acquire the reliability of the designed system which is revealed in Table 7.3 and represented graphically in Figure 7.3.

Table 7.3 Reliability as function of time

Time (t)	Reliability $R(t)$
0	1.00000
1	0.93417
2	0.87089
3	0.81038
4	0.75276
5	0.69814
6	0.64652
7	0.59791
8	0.55225
9	0.50948
10	0.46950
11	0.43220
12	0.39749
13	0.36523
14	0.33530
15	0.30757
16	0.28192
17	0.25822
18	0.23636
19	0.21620
20	0.19764

7.3.3 Mean Time to Failure

MTTF is an average operating time of the system without failure during a particular function under specified conditions. Substituting all repair rates equal to zero and putting the limit s tends to zero in Equation (7.16); one can obtain the MTTF in terms of failure rates as

$$MTTF = \lim_{s \to 0} \overline{P}_{up}(s)$$
$$= \frac{(\lambda_{PS} + \lambda_M + \lambda_{MC} + \lambda_F + \lambda_{SPS} + \lambda_P)}{(\lambda_{PS} + \lambda_M + \lambda_{MC} + \lambda_F + \lambda_P)(\lambda_M + \lambda_{MC} + \lambda_F + \lambda_{SPS} + \lambda_P)}.$$
$$(7.21)$$

Now varying input parameters one by one at 0.1, 0.2, ..., 0.9, respectively, and setting the failure rate as $\lambda_{PS} = 0.045$, $\lambda_M = 0.012$, $\lambda_{MC} = 0.024$, $\lambda_F = 0.024$, $\lambda_{SPS} = 0.045$, $\lambda_P = 0.007$ in Equation (7.46), we can get the variation of MTTF regarding failure rates that displayed in Table 7.4 and graphical representation shown in Figure 7.4.

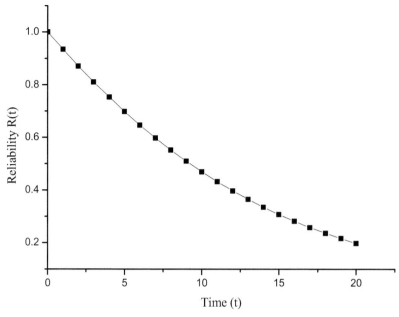

Figure 7.3 Reliability as function of time.

Table 7.4 MTTF as function of failure rates

Variation in $\lambda_{PS}, \lambda_M, \lambda_{MC}, \lambda_F, \lambda_{SPS}, \lambda_P$	MTTF with respect to failure rates					
	λ_{PS}	λ_M	λ_{MC}	λ_F	λ_{SPS}	λ_P
0.1	11.33447	6.09756	6.55936	6.55936	11.10599	5.92334
0.2	10.43339	3.82514	4.00550	4.00550	10.14757	3.75463
0.3	10.02336	2.77778	2.87243	2.87243	9.71145	2.74014
0.4	9.78893	2.17822	2.23622	2.23622	9.46211	2.15492
0.5	9.63719	1.79063	1.82974	1.82974	9.30071	1.77483
0.6	9.53095	1.51976	1.54787	1.54787	9.18771	1.50834
0.7	9.45241	1.31987	1.34105	1.34105	9.10418	1.31125
0.8	9.39199	1.16636	1.18287	1.18287	9.03992	1.15961
0.9	9.34407	1.04478	1.05801	1.05801	8.98894	1.03936

7.3.4 Sensitivity Analysis

Sensitivity analysis is the analysis that computes how the uncertainty in the output or sensitive an output of a mathematical model or a complex system can be allocate to various sources of uncertainty in its inputs or to change in an input while keeping other inputs constant.

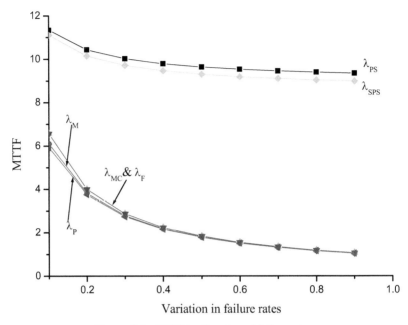

Figure 7.4   MTTF as function of failure rates.

7.3.4.1 Sensitivity of Reliability

Reliability sensitivity can be computed by differentiating partially condition (7.19) regarding to their input variable and fixing the values of input
parameters as $\lambda_{PS} = 0.045$, $\lambda_M = 0.012$, $\lambda_{MC} = 0.024$, $\lambda_F = 0.024$,
$\lambda_{SPS} = 0.045$, $\lambda_P = 0.007$ in the partial derivatives of Equation (7.19), we
get Table 7.5 and correspondingly Figure 7.5.

7.3.4.2 Sensitivity of MTTF

MTTF sensitivity can be estimate by partial differentiation of Equation (7.21)
with respect to the input measures and then setting the values of input
parameters as $\lambda_{PS} = 0.045$, $\lambda_M = 0.012$, $\lambda_{MC} = 0.024$, $\lambda_F = 0.024$,
$\lambda_{SPS} = 0.045$, $\lambda_P = 0.007$ in partial derivatives of Equation (7.21), one
may get Table 7.6 and corresponding Figure 7.6.

7.4 Results Conversation

The proposed work investigates the availability, reliability, MTTF, and sensitivity of reliability and MTTF of the designed VTOL prototype system

Table 7.5 Reliability sensitivity as function of time

Time (t)	Reliability sensitivity					
	$\dfrac{\partial R(t)}{\partial \lambda_{PS}}$	$\dfrac{\partial R(t)}{\partial \lambda_{M}}$	$\dfrac{\partial R(t)}{\partial \lambda_{MC}}$	$\dfrac{\partial R(t)}{\partial \lambda_{F}}$	$\dfrac{\partial R(t)}{\partial \lambda_{SPS}}$	$\dfrac{\partial R(t)}{\partial \lambda_{P}}$
0	0.00000	0.00000	0.00000	0.00000	0.00000	0.00000
1	−0.02005	−0.93417	−0.93417	−0.93417	−0.02231	−0.93417
2	−0.07146	−1.74179	−1.74179	−1.74179	−0.07967	−1.74179
3	−0.14327	−2.43113	−2.43113	−2.43113	−0.15999	−2.43113
4	−0.22696	−3.01106	−3.01106	−3.01106	−0.25387	−3.01106
5	−0.31600	−3.49069	−3.49069	−3.49069	−0.35405	−3.49069
6	−0.40547	−3.87914	−3.87914	−3.87914	−0.45505	−3.87914
7	−0.49178	−4.18537	−4.18537	−4.18537	−0.55284	−4.18537
8	−0.57236	−4.41801	−4.41801	−4.41801	−0.64450	−4.41801
9	−0.64550	−4.58528	−4.58528	−4.58528	−0.72806	−4.58528
10	−0.71011	−4.69495	−4.69495	−4.69495	−0.80227	−4.69495
11	−0.76565	−4.75426	−4.75426	−4.75426	−0.86646	−4.75426
12	−0.81194	−4.76989	−4.76989	−4.76989	−0.92038	−4.76989
13	−0.84912	−4.74801	−4.74801	−4.74801	−0.96413	−4.74801
14	−0.87752	−4.69422	−4.69422	−4.69422	−0.99804	−4.69422
15	−0.89764	−4.61362	−4.61362	−4.61362	−1.02263	−4.61362
16	−0.91008	−4.51079	−4.51079	−4.51079	−1.03853	−4.51079
17	−0.91551	−4.38984	−4.38984	−4.38984	−1.04646	−4.38984
18	−0.91460	−4.25442	−4.25442	−4.25442	−1.04717	−4.25442
19	−0.90806	−4.10776	−4.10776	−4.10776	−1.04142	−4.10776
20	−0.89658	−3.95273	−3.95273	−3.95273	−1.02997	−3.95273

with the help of stochastic modeling. The effects of the above findings are discussed further.

Figure 7.2 shows the changes in the availability of the system. From the critical examination of the availability graph, one can see that availability of the system reduce sharply in a short period, but after a time period, it reduce slightly as time increases. It seems to be constant over a long period.

Figure 7.3 gives an idea about the reliability of the considered VTOL system. The reliability graph of the system clearly reveals that how reliable the system. Reliability of the VTOL system reduces smoothly in a constant manner as time passes.

In Figure 7.4, the authors studied the MTTF regarding to the variation in failure rates. Vital examination of Figure 7.4 revealed that the MTTF of the system decreases with respect to the failure rate of each part of the VTOL system. In the case of variation in the failure rate of the motor controller and

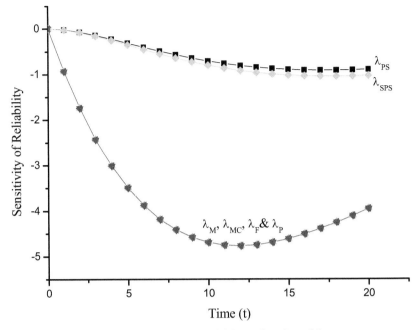

Figure 7.5 Reliability sensitivity as function of time.

fight controller, the system has identical MTTF. After some variation in the failure rate of each unit except the power supply and standby, the value of system's MTTF coincides with each other.

Figure 7.5 shows the trend of the sensitivity of the system reliability and one can observe that the sensitivity of reliability decreases as time passes with respect to the failure rate of the power supply and standby. In concern of the failure rate of the motor, motor controller, flight controller, and propeller, the sensitivity of the system reliability decreases smoothly first and tends to increase after a long run. With respect to these failure rates, sensitivity reliability coincides with each other. It is clear from the graph that the system reliability is more sensitive regarding to the power supply. Figure 7.6 yields the sensitivity of system's MTTF and shows that the sensitivity of system's MTTF increases with respect to each failure rate as failure rate increases. System's MTTF is equally sensitive in case of the motor controller and flight. After some increases in the failure rate of the motor, motor controller, flight, and propeller, the sensitivity of the system's MTTF coincide with each other. Similarly, after a bit increment in the failure rate of the power supply and standby, the sensitivity of the system's MTTF coincide with each other. It is

Table 7.6 MTTF sensitivity as function of failure rates

Variation in $\lambda_{PS}, \lambda_M, \lambda_{MC}, \lambda_F, \lambda_{SPS}, \lambda_P$	MTTF sensitivity with respect to failure rates					
	$\dfrac{\partial MTTF}{\partial \lambda_{PS}}$	$\dfrac{\partial MTTF}{\partial \lambda_M}$	$\dfrac{\partial MTTF}{\partial \lambda_{MC}}$	$\dfrac{\partial MTTF}{\partial \lambda_F}$	$\dfrac{\partial MTTF}{\partial \lambda_{SPS}}$	$\dfrac{\partial MTTF}{\partial \lambda_P}$
0.1	−14.40660	−35.84176	−41.31623	−41.31623	−15.32326	−33.87197
0.2	−5.63601	−14.36292	−15.72804	−15.72804	−5.99462	−13.84554
0.3	−2.98306	−7.63032	−8.15410	−8.15410	−3.17287	−7.42676
0.4	−1.84230	−4.70934	−4.96183	−4.96183	−1.95952	−4.60976
0.5	−1.24976	−3.18929	−3.32943	−3.32943	−1.32928	−3.13349
0.6	−0.90311	−2.30042	−2.38601	−2.38601	−0.96058	−2.26611
0.7	−0.68297	−1.73665	−1.79265	−1.79265	−0.72643	−1.71408
0.8	−0.53451	−1.35700	−1.39561	−1.39561	−0.56852	−1.34139
0.9	−0.42968	−1.08933	−1.11705	−1.11705	−0.45702	−1.07808

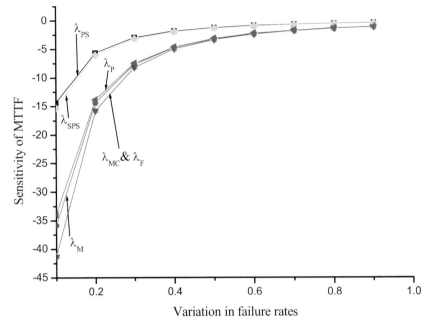

Figure 7.6   MTTF sensitivity as function of failure rates.

clear from the graph that the system's MTTF is more sensitive regarding to the power supply and standby.

7.5 Conclusion

VTOL quadrotor systems have become progressively platforms and essential for our economic competitiveness. From this research work, it is concluded that the availability, reliability, and MTTF of the VTOL system decreases; and the system is highly sensitive with respect to the failure rate of the power supply and standby. By controlling these failure rates, one can improve system's MTTF and can attain a highly reliable system. In the future, it is expected that the proposed reliability modeling can seek more applications in reliability engineering of VTOL systems. It is also very useful in such fields as Ariel surveillance in scenarios of war and natural calamities and epidemics, Ariel is filming and Ariel cinematography, capturing videos, and photos of hard to reach places and for recreational purposes.

References

Avizienis, A., Laprie, J. C., Randell, B., and Landwehr, C. (2004). Basic concepts and taxonomy of dependable and secure computing. IEEE Transactions on Dependable and Secure Computing, 1(1), 11–33.

Berbra, C., Lesecq, S., Gentil, S., and Thiriet, J. M. (2008). Co-design of a safe network control quadrotor. In 17th IFAC World Congress (pp. 5506–5511).

Bouabdallah, S., and Siegwart, R. (2007). Full control of a quadrotor. In Intelligent robots and systems, 2007. IROS 2007. IEEE/RSJ international conference on (pp. 153–158). IEEE.

Cui, L., and Li, H. (2007). Analytical method for reliability and MTTF assessment of coherent systems with dependent components. Reliability Engineering and System Safety, 92(3), 300–307.

Goyal, N., Ram, M., and Mittal, A. (2015). Reliability measures analysis of a computer system incorporating two types of repair under copula approach. In Numerical Methods for Reliability and Safety Assessment, (pp. 365–385). Springer International Publishing.

Goyal, N., Sharma, P., and Ram, M. (2015). Automated teller machine performance evaluation through cash transactions. Mathematics in Engineering, Science and Aerospace (MESA), 6(2).

Goyal, N., Kaushik, A., and Ram, M. (2016). Automotive water cooling system analysis subject to time dependence and failure issues. International Journal of Manufacturing, Materials, and Mechanical Engineering (IJMMME), 6(2), 1–22.

Goyal, N., Ram, M., Bhardwaj, A., and Kumar, A. (2016). Thermal power plant modelling with fault coverage stochastically. International Journal of Manufacturing, Materials, and Mechanical Engineering (IJMMME), 6(3), 28–44.

Goyal, N., Ram, M., and Kaushik, A. (2017). Performability of solar thermal power plant under reliability characteristics. International Journal of System Assurance Engineering and Management, 1–9.

Goyal, N., and Ram, M. (2017). Series-parallel system study under warranty and preventive maintenance. In Mathematics Applied to Engineering, (pp. 97–113).

Hoffmann, G. M., Huang, H., Waslander, S. L., and Tomlin, C. J. (2007). Quadrotor helicopter flight dynamics and control: Theory and experiment. In Proc. of the AIAA Guidance, Navigation, and Control Conference (Vol. 2).

Kumar, K., Singh J., and Kumar P. (2009). Fuzzy reliability and fuzzy availability of the serial process in butter-oil processing plant. Journal of Mathematics and Statistics, 5(1), 65–71.

Manglik, M., and Ram, M. (2013). Reliability analysis of a two unit cold standby system using Markov process. Journal of Reliability and Statistical Studies, 6(2).

Olson, I. J., and Atkins, E. M. (2013). Qualitative failure analysis for a small quadrotor unmanned aircraft system. AIAA Guidance, Navigation, and Control (GNC) Conference. doi: 10.2514/6.2013-4761.

Ram, M., and Singh, S. B. (2010). Analysis of a complex system with common cause failure and two types of repair facilities with different distributions in failure. International Journal of Reliability and Safety, 4(4), 381–392.

Ram, M., and Kumar, A. (2015). Performability analysis of a system under 1-out-of-2: G scheme with perfect reworking. Journal of the Brazilian Society of Mechanical Sciences and Engineering, 37(3), 1029–1038.

Ram, M., Nagiya, K., and Goyal, N. (2015). Mathematical modelling with an application to nuclear power plant reliability analysis. In Research Advances in Industrial Engineering, (pp. 89–102), Springer International Publishing.

Ram, M., and Goyal, N. (2015). Gas turbine assimilation under copula-coverage approaches. In Research Advances in Industrial Engineering, (pp. 103–116). Springer International Publishing.

Ram, M., and Davim, J. P. (Eds.). (2017). Advances in Reliability and System Engineering. Springer International Publishing.

Sa, I., and Corke, P. (2011). Estimation and control for an open-source quadcopter. In Proceedings of the Australasian Conference on Robotics and Automation 2011.

Index

About the Authors

Dr. Kanchan Das is an Associate Professor at East Carolina University, North Carolina, USA. He received his PhD in Industrial Engineering from the University of Windsor, Canada. He is a member of Institute of Industrial and Systems Engineers and Decision Sciences Institute. His research interests include Integrating Resilience in Supply Chain Network Design and Planning; Design and Planning of Sustainable Supply Chain; Integrating Lean Systems and Tools for improving Supply Chain Sustainability; Reliability and Sustainability Considerations in the Design and Planning of Manufacturing Systems. He is an Editorial Board Member for International Journal of Mathematical, Engineering and Management Sciences.

Dr. Mangey Ram received the PhD degree major in Mathematics and minor in Computer Science from G. B. Pant University of Agriculture and Technology, Pantnagar, India. He has been a Faculty Member for around 10 years and has taught several core courses in pure and applied mathematics at undergraduate, postgraduate, and doctorate levels. He is currently a Professor at Graphic Era (Deemed to be University), Dehradun, India. Before joining the Graphic Era, he was a Deputy Manager (Probationary Officer) with Syndicate Bank for a short period. He is Editor-in-Chief of *International Journal of Mathematical, Engineering and Management Sciences* and the Guest Editor & Member of the editorial board of various journals. He is a regular reviewer for international journals, including IEEE, Elsevier, Springer, Emerald, John Wiley, Taylor & Francis, and many other publishers. He has published 144 research publications in IEEE, Taylor & Francis, Springer, Elsevier, Emerald, World Scientific and many other national and international journals of repute and also presented his works at national and international conferences. His fields of research are reliability theory and applied mathematics. Dr. Ram is a Senior Member of the IEEE, life member of Operational Research Society of India, Society for Reliability Engineering, Quality and Operations Management in India, Indian Society of Industrial and Applied Mathematics, member of International Association of Engineers in Hong Kong, and Emerald Literati Network in the UK. He has been a member of the organizing committee of a number of international and national conferences, seminars, and workshops. He has been conferred with *Young Scientist Award* by the Uttarakhand State Council for Science and Technology, Dehradun, in 2009. He has been awarded the *Best Faculty Award* in 2011; *Research Excellence Award* in 2015; and recently *Outstanding Researcher Award* in 2018 for his significant contribution in academics and research at Graphic Era (Deemed to be University, Dehradun, India).